# What Your Colleagues Are Saying . . .

"Wow! This book is a treasure trove for middle school teachers and those who support middle level mathematics education. The tasks presented are top notch, the student work is illustrative of classrooms, and the suggestions for how to respond are incredible! Each of these three would make a great resource—to have them all blended together is outstanding."

—Kevin Dykema, Math Teacher,
Mattawan Middle School, Mattawan, MI

"*Mine the Gap for Mathematical Understanding* provides the framework and guidance teachers need to move beyond finding correct answers to dig deeply into the mine of student thinking, to analyze misconceptions and gaps in understanding, and to develop and implement specific strategies to support every student in deep mathematics learning. Every teacher of mathematics needs this amazing resource to learn to mine the gaps for each of their students!"

—Becky M. Walker, Director of Learning Services,
Cooperative Educational Services Agency #7, Green Bay, WI

"This work does what other books only attempt to do. It combines instruction, assessment, and practice with open-ended and rich tasks that allow for teachers to not only immediately implement the ideas but also understand the content and pedagogy behind them. The tasks, which are immediately implementable and customizable, engage each and every learner. They are based on cutting-edge and research-based instructional frameworks and provide countless learning opportunities for students."

—Zachary Champagne, Assistant in Research, Florida Center for Research in Science,
Technology, Engineering, and Mathematics at Florida State University

"*Mine the Gap* is a great tool for teachers to use to grow their own understanding of student misconceptions and incomplete understandings and how to address them. This is an indispensable resource for all involved in supporting students' growth in mathematics."

—Nathan Rosin, Sun Prairie Area School District

"More than just a nice collection of problems, this book shares a road map for teachers looking to enhance the quality of the math tasks they use with students. Teachers will appreciate the examples of actual student work paired with tips for analysis and instruction."

—Delise Andrews, Mathematics Coordinator, Lincoln (NE) Public Schools

"This book helps navigate how to use student work to drive instruction with rich engaging tasks, which will help all students become better mathematicians. The authors have done an excellent job of helping teachers to carefully look at student work to identify how students solved math problems, using this evidence to identify those students who understand the targeted skill, along with the misconceptions or misunderstandings of other students, with suggestions of how to move all students forward in their thinking."

—Cynthia Baumann, Mathematics Supervisor, Omaha Public Schools

*For my countless math friends who encourage and support me. You know who you are. Every conversation makes me better. A special few took a chance on me. They gave me opportunities. I am forever grateful. This is also for my favorite middle school math teacher and best friend of all.*

*—John*

*For my two favorite guys, Nick and Justin. To all of my family and friends who have supported me over the years. For my Granny, who inspired my love for learning. I miss you every day.*

*—Jenny*

# MINE ᴛʜᴇ GAP
## FOR MATHEMATICAL UNDERSTANDING

**Common Holes and Misconceptions
and What To Do About Them**

John SanGiovanni

Jennifer Rose Novak

**GRADES 6–8**

**FOR INFORMATION:**

Corwin

A SAGE Company

2455 Teller Road

Thousand Oaks, California 91320

(800) 233-9936

www.corwin.com

SAGE Publications Ltd.

1 Oliver's Yard

55 City Road

London, EC1Y 1SP

United Kingdom

SAGE Publications India Pvt. Ltd.

B 1/I 1 Mohan Cooperative Industrial Area

Mathura Road, New Delhi 110 044

India

SAGE Publications Asia-Pacific Pte. Ltd.

3 Church Street

#10-04 Samsung Hub

Singapore 049483

Program Manager, Mathematics:  Erin Null

Editorial Development Manager:  Julie Nemer

Editorial Assistant:  Nicole Shade

Production Editor:  Melanie Birdsall

Copy Editor:  Liann Lech

Typesetter:  Integra

Proofreader:  Susan Schon

Cover and Interior Designer:  Scott Van Atta

Marketing Manager:  Margaret O'Connor

Printed in the United States of America

*Library of Congress Cataloging-in-Publication Data*

Names: SanGiovanni, John, author. | Novak, Jennifer Rose, author.

Title: Mine the gap for mathematical understanding, grades 6–8 : common holes and misconceptions and what to do about them / John SanGiovanni, Jennifer Rose Novak.

Description: Thousand Oaks, California : Corwin, [2018] | Includes bibliographical references.

Identifiers: LCCN 2017022721 | ISBN 9781506379821 (pbk.)

Subjects: LCSH: Mathematics—Study and teaching (Middle school) | Middle school teachers—Training of. | Mathematics teachers—Training of.

Classification: LCC QA135.6 .S2584 2018 | DDC 510.71/2—dc23 LC record available at https://lccn.loc.gov/2017022721

This book is printed on acid-free paper.

21 1 0 9 8 7 6 5 4 3

# MINE the GAP

## FOR MATHEMATICAL UNDERSTANDING

A quick-reference matrix provides a snapshot of the Big Ideas in the book, along with descriptions of the associated tasks.

## BIG IDEAS AND TASKS AT A GLANCE

| Big Idea No. | Big Idea | Task | Description |
|---|---|---|---|
| 1 | Addition and Subtraction of Fractions | 1A | Students add mixed numbers on a number line to justify their sums. |
| 1 | Addition and Subtraction of Fractions | 1B | Students find different fractions with the same sum. |
| 1 | Addition and Subtraction of Fractions | 1C | Students consider subtraction misconceptions. |
| 1 | Addition and Subtraction of Fractions | 1D | Students add and subtract fractions to solve a problem. |
| 2 | Multiplication and Division of Fractions | 2A | Students decompose mixed numbers to multiply. |
| 2 | Multiplication and Division of Fractions | 2B | Students consider misconceptions about multiplication of fractions. |
| 2 | Multiplication and Division of Fractions | 2C | Students solve problems with multiplication and division of fractions. |
| 2 | Multiplication and Division of Fractions | 2D | Students consider the results of dividing with fractions. |
| 3 | Reasoning About Addition and Subtraction of Fractions | 3A | Students reason about the difference of fractions. |
| 3 | Reasoning About Addition and Subtraction of Fractions | 3B | Students compare the sums of fractions. |
| 3 | Reasoning About Addition and Subtraction of Fractions | 3C | Students compare sums of fractions to a benchmark. |
| 3 | Reasoning About Addition and Subtraction of Fractions | 3D | Students reason about subtraction with mixed numbers. |

# CHAPTER 4

## RATIO, PROPORTION, AND PERCENT

**THIS CHAPTER HIGHLIGHTS HIGH-QUALITY TASKS FOR THE FOLLOWING:**

- Big Idea 17: Representing Ratios

  Ratios can be represented in different ways. But ratios can also represent different ideas, including part-to-part and part-to-whole relationships.

- Big Idea 18: Equivalent Ratios

  Equivalent ratios model the same relationship between two quantities. We can use a variety of representations to model this equivalency.

- Big Idea 19: Unit Rates

  Finding and applying unit rates allows us to have a base of comparison between two scenarios. Unit rates also produce a scale factor for finding other equivalent ratios.

- Big Idea 20: Using Ratios to Solve Problems

  Ratio reasoning provides multiple strategies for solving real-world problems.

- Big Idea 21: Reasoning With Percents

  Tasks involving percent build on understanding of ratio reasoning, because percents are values compared to a whole of 100.

- Big Idea 22: Unit Rate as Slope

  Understanding unit rate as slope leads to understanding of linear functions. We can represent this relationship in different ways to determine the constant of proportionality.

*Chapter Overviews* highlight and explain the Big Ideas covered in each chapter.

The highlighted task is explained in depth and potential student responses are predicted and described in detail.

Each Big Idea starts by describing one related high-quality task.

---

**BIG IDEA 18**
## Equivalent Ratios

### TASK 18A

> Jenny is sure that 10:12 is equivalent to 25:30. Do you agree with Jenny?
>
> Use pictures, numbers, or words to justify your thinking.

 **MODIFYING THE TASK**

We can modify the numbers used in the ratio relative to our students' understanding and number sense development.

**About the Task**

Ideas about equivalent ratio extend well beyond the ability to calculate equivalency. Representations support students' understanding of the concept of equivalent ratios. *The open-ended nature of this problem allows the student to select his or her own representation and interpretation of each ratio.* The temptation to calculate may be limited due to the nature of the numbers in the ratios. This is especially noteworthy because there may be problems or tasks with numbers that enable students to find convenient "solutions," possibly without deep understanding. We should also note that this task prompts the students to agree or disagree rather than correct flawed ideas. Our students then need to verify their thinking.

⚠ **MINING HAZARD**

Pictorial representations are appropriate ways to show relationships between two quantities, but using them to verify equivalence can be difficult as numbers or relationships become more complex.

**Anticipating Student Responses**

Students will typically represent the ratios using tape diagrams or double number line models, as these are the clearest way to show equivalence. *Some students may illustrate these ratios with pictorial representations. Some students may rely on contexts to make sense of ratios. Yet in this task, these*

*Mining Hazard* icons signal examples of incomplete thinking that students may encounter.

*Pause and Reflect* sections invite teachers to think about the task in relation to their practice and their own students.

### PAUSE AND REFLECT

- How does this task compare to tasks I've used?
- What might my students do in this task?

 Visit this book's companion website at **resources.corwin.com/minethegap/6-8** for complete, downloadable versions of all tasks.

### Student 1

Student 1's work shows significant misunderstanding. Her work represents two different challenges students might have when working with this problem. *She subtracts $2\frac{1}{2}$ from 10 because she notes that the problem asks "how many more" and she connects this phrase with subtraction.* Her computation is also flawed. She says $\frac{10}{1} - 2\frac{1}{2} = 2\frac{9}{1}$. To make sense of this error, we can presume that she subtracts in either direction to find a result. In other words, she subtracted left to right for the numerator (10 − 1). She subtracts right to left for the denominator (2 − 1). It's likely that she simply brings the whole number over to her "difference."

### Student 2

Student 2 uses the more efficient strategy. She finds the difference of the two days (9 − 5). She then multiplies the difference by $2\frac{1}{2}$. She uses this strategy instead of multiplying both values by $2\frac{1}{2}$ and then finding the difference of the products. Her note about the "key" communicates that each number in the table is multiplied by $2\frac{1}{2}$. She doesn't multiply $4 \times 2\frac{1}{2}$ correctly. Instead, she multiplies 4 × 2 and then adds the half to the product yielding an inaccurate result of $8\frac{1}{2}$.

*What would we want to ask these students? What might we do next?*

### Student 1

This problem shows that Student 1 needs work with problem solving and computation. We can work on both concepts at the same time. First, we want to develop problem-solving strategies. We can have her work with models and drawings and connect these with equations. Computation work should revisit addition and subtraction of fractions. We would be wise to begin with fractions less than 1 before moving to mixed numbers. It would also be wise to work with less complicated denominators, such as halves, fourths, eighths, and twelfths.

### Student 2

Student 2 shows proficiency with solving the problem. She has made sense of the problem and applies an efficient strategy. *Her written results follow whe she had subtracted 9 − 5.* Looking _____ associated days. But after _____ which is not relevant to th_____ with fractions and mixed n_____ include the half in her multi_____ problem.

**MINING TIP**

Avoid using key words as an instructional approach to problem solving. Doing so can create misconceptions about problems and set students up for incorrect solutions.

**MINING HAZARD**

We must consider all of a student's work when determining his or her understanding. We can combine students' written thoughts with their diagrams or drawings to establish full understanding. However, additional ideas may not always link their ideas.

---

**TASK 4A:** Which expression has the greatest quotient? Use models, numbers, or words to explain your thinking.

$$20 \div \frac{1}{4} \qquad 20 \div \frac{1}{2} \qquad 20 \div \frac{1}{5}$$

**Student Work 3**

**Student Work 4**

---

*What They Did* sections analyze how the students' work gives insight into their thinking.

*Mining Tip* icons offer additional notes about mathematics content, misconceptions, or implementing the related tasks.

*Using Evidence* sections identify questions and instructional next steps to address gaps in student understanding.

*Mining Hazard* icons also offer insight and advice as to where teachers themselves sometimes go awry in their own thinking.

Each task is highlighted at the top of the page, with the related student work showcased below.

OTHER TASKS

- What will count as evidence of understanding?
- What misconceptions might you find?
- What will you do or how will you respond?

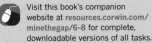
Visit this book's companion website at resources.corwin.com/minethegap/6-8 for complete, downloadable versions of all tasks.

**TASK 4B:** Danny knows $50 \times \frac{1}{2} = 25$. Because of this, she knows that $50 \times \frac{1}{4}$ must be less. Do you agree with Danny? What do you think the new product might be? Use pictures, numbers, or words to explain your thinking.

Understanding how the size of factors impact products is critical for determining the reasonableness of our answers. It is likely that some, if not most or even all, of our students believe that multiplication yields a product larger than the first factor, but as we know, this isn't the case. In this task, our students are asked to describe why a product of a factor and $\frac{1}{4}$ will be less than the product of the same factor and $\frac{1}{2}$. Our students are likely to reason that it makes sense because $\frac{1}{4}$ is less than $\frac{1}{2}$. It will be interesting to see if they recognize that the product will be 12.5 (or exactly half of 25). Some students may complete the computation to justify their solution. This may not be indicative of the reasoning we are seeking in our students.

**TASK 4C:** Oscar noticed these two columns of equations on the board.

$50 \div \frac{1}{10} = 500$          $50 \div \frac{2}{10} = 250$

$40 \div \frac{1}{10} = 400$          $40 \div \frac{2}{10} = 200$

$30 \div \frac{1}{10} = 300$          $30 \div \frac{2}{10} = 150$

$20 \div \frac{1}{10} = 200$          $20 \div \frac{2}{10} = 100$

**What patterns do you notice about the equations?**

**So how can knowing $80 \div \frac{1}{10} = 800$ help you solve $80 \div \frac{2}{10}$ ?**

*Intentional arrangement of equations can help our students see patterns in computations.* The quotients of the equations are provided so that students can focus on the relationships. We want our students to notice that the quotients of a number and $\frac{2}{10}$ are half of the quotient of the number and $\frac{1}{10}$. Though students may recognize the pattern, they may not be able to explain that they are dividing by twice as much so then the quotient will be half as much. The extension on the task is an opportunity to show that students can generalize their understanding.

MODIFYING THE TASK

This task can be modified to examine patterns when computing other numbers. For example, we could explore the results of multiplying by $\frac{1}{10}$ and $\frac{2}{10}$ by simply changing the operation in the prompt.

*Other Tasks* sections provide three additional high-quality tasks related to each Big Idea, along with relevant explanations and analyses.

*Modifying the Task* marginal notes provide suggestions for further adaptation and exploration.

# CONTENTS

# BIG IDEAS AND TASKS AT A GLANCE

| Big Idea No. | Big Idea | Task | Description |
|---|---|---|---|
| 1 | Addition and Subtraction of Fractions | 1A | Students add mixed numbers on a number line to justify their sums. |
| 1 | Addition and Subtraction of Fractions | 1B | Students find different fractions with the same sum. |
| 1 | Addition and Subtraction of Fractions | 1C | Students consider subtraction misconceptions. |
| 1 | Addition and Subtraction of Fractions | 1D | Students add and subtract fractions to solve a problem. |
| 2 | Multiplication and Division of Fractions | 2A | Students decompose mixed numbers to multiply. |
| 2 | Multiplication and Division of Fractions | 2B | Students consider misconceptions about multiplication of fractions. |
| 2 | Multiplication and Division of Fractions | 2C | Students solve problems with multiplication and division of fractions. |
| 2 | Multiplication and Division of Fractions | 2D | Students consider the results of dividing with fractions. |
| 3 | Reasoning About Addition and Subtraction of Fractions | 3A | Students reason about the difference of fractions. |
| 3 | Reasoning About Addition and Subtraction of Fractions | 3B | Students compare the sums of fractions. |
| 3 | Reasoning About Addition and Subtraction of Fractions | 3C | Students compare sums of fractions to a benchmark. |
| 3 | Reasoning About Addition and Subtraction of Fractions | 3D | Students reason about subtraction with mixed numbers. |

| Big Idea No. | Big Idea | Task | Description |
|---|---|---|---|
| 4 | Reasoning About Multiplication and Division of Fractions | 4A | Students interpret quotients of fractions. |
| 4 | Reasoning About Multiplication and Division of Fractions | 4B | Students reason about products of fractions. |
| 4 | Reasoning About Multiplication and Division of Fractions | 4C | Students observe patterns of multiplication with fractions. |
| 4 | Reasoning About Multiplication and Division of Fractions | 4D | Students reason about quotients of fractions. |
| 5 | Problem Solving With Fractions | 5A | Students solve problems with fractions in a table. |
| 5 | Problem Solving With Fractions | 5B | Students solve a multi-step problem. |
| 5 | Problem Solving With Fractions | 5C | Students solve an open-ended problem. |
| 5 | Problem Solving With Fractions | 5D | Students represent and write a problem for division with fractions. |
| 6 | Decimals as Numbers | 6A | Students consider decimals on a number line. |
| 6 | Decimals as Numbers | 6B | Students represent decimals on a number line. |
| 6 | Decimals as Numbers | 6C | Students relate decimals through a decimal chart. |
| 6 | Decimals as Numbers | 6D | Students decompose decimals. |
| 7 | Addition and Subtraction With Decimals | 7A | Students reason about the sums of decimals. |
| 7 | Addition and Subtraction With Decimals | 7B | Students add and subtract decimals on a number line. |
| 7 | Addition and Subtraction With Decimals | 7C | Students reason about adding and subtracting with decimals. |
| 7 | Addition and Subtraction With Decimals | 7D | Students reason about subtraction with decimals to place a decimal point. |
| 8 | Multiplication and Division With Decimals | 8A | Students reason about quotients of decimals. |
| 8 | Multiplication and Division With Decimals | 8B | Students interpret quotients of decimals. |
| 8 | Multiplication and Division With Decimals | 8C | Students represent multiplication of decimals with an area model. |

| Big Idea No. | Big Idea | Task | Description |
|---|---|---|---|
| 8 | Multiplication and Division With Decimals | 8D | Students compare products of decimals by reasoning. |
| 9 | Representing Integers | 9A | Students decompose integers. |
| 9 | Representing Integers | 9B | Students represent an integer in four different ways. |
| 9 | Representing Integers | 9C | Students consider integer relationships on an integer chart. |
| 9 | Representing Integers | 9D | Students identify real-world examples of integers. |
| 10 | Representing Integers on Number Lines | 10A | Students reason about where to place numbers on an open number line. |
| 10 | Representing Integers on Number Lines | 10B | Students consider relationships between integers on number lines. |
| 10 | Representing Integers on Number Lines | 10C | Students consider the changing position of an integer on different number lines. |
| 10 | Representing Integers on Number Lines | 10D | Students represent the same integer on different number lines. |
| 11 | More Representing Integers on Number Lines | 11A | Students reason about how an integer relates to other integers on two different number lines. |
| 11 | More Representing Integers on Number Lines | 11B | Students consider midpoints of number lines with different integers. |
| 11 | More Representing Integers on Number Lines | 11C | Students consider integers on number lines with different intervals. |
| 11 | More Representing Integers on Number Lines | 11D | Students consider how endpoints change the value of a point between them. |
| 12 | Comparing Integers | 12A | Students react to misconceptions about integer comparison. |
| 12 | Comparing Integers | 12B | Students use number lines to compare integers. |
| 12 | Comparing Integers | 12C | Students create integers to compare to a given benchmark. |
| 12 | Comparing Integers | 12D | Students order a collection of integers. |
| 13 | Addition With Integers | 13A | Students reason about addition with integers. |
| 13 | Addition With Integers | 13B | Students add integers on a number line. |
| 13 | Addition With Integers | 13C | Students decompose integers to add more efficiently. |
| 13 | Addition With Integers | 13D | Students add integers with an integer chart. |

| Big Idea No. | Big Idea | Task | Description |
|---|---|---|---|
| 14 | Subtraction With Integers | 14A | Students create pairs of integers for a given difference. |
| 14 | Subtraction With Integers | 14B | Students represent subtraction with integers. |
| 14 | Subtraction With Integers | 14C | Students find a specific difference of different integers. |
| 14 | Subtraction With Integers | 14D | Students estimate differences of integers. |
| 15 | Multiplication With Integers | 15A | Students compare products of integers. |
| 15 | Multiplication With Integers | 15B | Students examine patterns within multiplication equations. |
| 15 | Multiplication With Integers | 15C | Students examine patterns with multiplication of integers on a number line. |
| 15 | Multiplication With Integers | 15D | Students reason about the relationship between integer factors. |
| 16 | Division With Integers | 16A | Students examine patterns with division of integers on a number line. |
| 16 | Division With Integers | 16B | Students represent division of integers in different ways. |
| 16 | Division With Integers | 16C | Students use relationships between division expressions to find quotients. |
| 16 | Division With Integers | 16D | Students examine patterns within division equations. |
| 17 | Representing Ratios | 17A | Students consider different representations of ratios. |
| 17 | Representing Ratios | 17B | Students represent ratio in different ways. |
| 17 | Representing Ratios | 17C | Students consider different ratios for the same collections. |
| 17 | Representing Ratios | 17D | Students work with real-world contexts for ratios. |
| 18 | Equivalent Ratios | 18A | Students justify why ratios are equivalent. |
| 18 | Equivalent Ratios | 18B | Students find equivalent ratios of different representations. |
| 18 | Equivalent Ratios | 18C | Students create and justify equivalent ratios. |
| 18 | Equivalent Ratios | 18D | Students solve real-world problems with equivalent ratios. |
| 19 | Unit Rates | 19A | Students apply unit rates to a real-world problem. |
| 19 | Unit Rates | 19B | Students identify unit rates. |

| Big Idea No. | Big Idea | Task | Description |
|---|---|---|---|
| 19 | Unit Rates | 19C | Students describe when we use unit rates. |
| 19 | Unit Rates | 19D | Students consider different interpretations of unit rates. |
| 20 | Using Ratios to Solve Problems | 20A | Students use ratio to solve problems. |
| 20 | Using Ratios to Solve Problems | 20B | Students use ratio to compare situations. |
| 20 | Using Ratios to Solve Problems | 20C | Students reason about ratio to compare situations. |
| 20 | Using Ratios to Solve Problems | 20D | Students use ratio to solve a real-world problem. |
| 21 | Reasoning With Percents | 21A | Students compare percents of quantities. |
| 21 | Reasoning With Percents | 21B | Students compare quantities to a benchmark percent. |
| 21 | Reasoning With Percents | 21C | Students reason about percents and quantity. |
| 21 | Reasoning With Percents | 21D | Students use percent to solve real-world problems. |
| 22 | Unit Rate as Slope | 22A | Students use unit rate as slope to solve a real-world problem. |
| 22 | Unit Rate as Slope | 22B | Students use unit rates and constants to solve a problem. |
| 22 | Unit Rate as Slope | 22C | Students use a unit rate to solve a problem. |
| 22 | Unit Rate as Slope | 22D | Students interpret graphs of unit rates. |
| 23 | Writing Expressions | 23A | Students consider expressions for the same situation. |
| 23 | Writing Expressions | 23B | Students create expressions for perimeter and area. |
| 23 | Writing Expressions | 23C | Students write an expression to solve a real-world problem. |
| 23 | Writing Expressions | 23D | Students create situations for given expressions. |
| 24 | Evaluating Expressions | 24A | Students reason about evaluations of expressions. |
| 24 | Evaluating Expressions | 24B | Students create and evaluate expressions. |
| 24 | Evaluating Expressions | 24C | Students evaluate an expression and generate new expressions with the same value. |
| 24 | Evaluating Expressions | 24D | Students consider different values for the same expression. |

| Big Idea No. | Big Idea | Task | Description |
|---|---|---|---|
| 25 | Equivalent Expressions | 25A | Students identify equivalent expressions. |
| 25 | Equivalent Expressions | 25B | Students create two expressions equivalent to a third expression. |
| 25 | Equivalent Expressions | 25C | Students use expressions to solve problems. |
| 25 | Equivalent Expressions | 25D | Students consider misconceptions about expressions. |
| 26 | Writing Equations | 26A | Students write an equation to model a problem. |
| 26 | Writing Equations | 26B | Students write an equation to model a problem. |
| 26 | Writing Equations | 26C | Students create a scenario for an equation. |
| 26 | Writing Equations | 26D | Students create different expressions to model a problem. |
| 27 | Solving Equations | 27A | Students consider errors when solving equations. |
| 27 | Solving Equations | 27B | Students create and solve an equation for a problem. |
| 27 | Solving Equations | 27C | Students identify equations with the same solution. |
| 27 | Solving Equations | 27D | Students consider if two equations have the same solution. |
| 28 | Inequalities | 28A | Students use inequalities with a real-world problem. |
| 28 | Inequalities | 28B | Students use inequalities with a real-world problem. |
| 28 | Inequalities | 28C | Students use an inequality to solve a problem. |
| 28 | Inequalities | 28D | Students compare solutions for inequalities. |
| 29 | Function Tables | 29A | Students describe and use a function table. |
| 29 | Function Tables | 29B | Students identify functional relationships. |
| 29 | Function Tables | 29C | Students find values for different functions. |
| 29 | Function Tables | 29D | Students consider functions in real-world contexts. |
| 30 | Reasoning About Graphing | 30A | Students reason about graphs. |
| 30 | Reasoning About Graphing | 30B | Students create a scenario for a graph. |
| 30 | Reasoning About Graphing | 30C | Students model a problem with a graph. |
| 30 | Reasoning About Graphing | 30D | Students compare the graphs of two functions. |

| Big Idea No. | Big Idea | Task | Description |
|---|---|---|---|
| 31 | Comparing Functions | 31A | Students compare different functions represented in different ways. |
| 31 | Comparing Functions | 31B | Students compare functions to solve a problem. |
| 31 | Comparing Functions | 31C | Students compare functions and consider the rate of change. |
| 31 | Comparing Functions | 31D | Students compare functions and consider the rate of change. |
| 32 | Systems of Equations | 32A | Students use systems of equations to create a new equation. |
| 32 | Systems of Equations | 32B | Students solve a real-world problem. |
| 32 | Systems of Equations | 32C | Students solve a real-world problem. |
| 32 | Systems of Equations | 32D | Students solve a system of equations and create a new equation. |
| 33 | Area of Composite Figures | 33A | Students find the area of a shape on a coordinate grid. |
| 33 | Area of Composite Figures | 33B | Students reason about the area of a figure. |
| 33 | Area of Composite Figures | 33C | Students create and find the area of a composite figure. |
| 33 | Area of Composite Figures | 33D | Students find the area of a composite figure with circles. |
| 34 | Nets and Three-Dimensional Figures | 34A | Students consider nets of rectangular prisms. |
| 34 | Nets and Three-Dimensional Figures | 34B | Students compare and contrast two nets. |
| 34 | Nets and Three-Dimensional Figures | 34C | Students consider an unconventional net. |
| 34 | Nets and Three-Dimensional Figures | 34D | Students create two different nets for a situation. |
| 35 | Surface Area and Volume | 35A | Students create prisms with the same volume. |
| 35 | Surface Area and Volume | 35B | Students use surface area to solve a problem. |
| 35 | Surface Area and Volume | 35C | Students find the dimensions of a triangular prism. |
| 35 | Surface Area and Volume | 35D | Students solve a real-world problem with surface area. |
| 36 | Volume of Cylinders and Cones | 36A | Students consider how dimensions change volume of cylinders. |

| Big Idea No. | Big Idea | Task | Description |
|---|---|---|---|
| 36 | Volume of Cylinders and Cones | 36B | Students find the dimensions of a cone. |
| 36 | Volume of Cylinders and Cones | 36C | Students compare cones and cylinders in a real-world context. |
| 36 | Volume of Cylinders and Cones | 36D | Students compare volume of cylinders and cones. |
| 37 | Angle Relationships | 37A | Students find angle measures. |
| 37 | Angle Relationships | 37B | Students determine possible angle measures for a triangle. |
| 37 | Angle Relationships | 37C | Students find exterior angles of triangles. |
| 37 | Angle Relationships | 37D | Students find angle measures of a figure with three parallel lines. |
| 38 | Transformations, Similarity, and Congruence | 38A | Students transform a figure on a coordinate plane. |
| 38 | Transformations, Similarity, and Congruence | 38B | Students describe a transformation on a coordinate plane. |
| 38 | Transformations, Similarity, and Congruence | 38C | Students transform a figure and describe the difference between the original and new image of the figure. |
| 38 | Transformations, Similarity, and Congruence | 38D | Students consider the results of a transformation. |
| 39 | Distance and the Pythagorean Theorem | 39A | Students find the lengths of segments of a figure. |
| 39 | Distance and the Pythagorean Theorem | 39B | Students consider if triangles are right triangles. |
| 39 | Distance and the Pythagorean Theorem | 39C | Students create right triangles from a given right triangle. |
| 39 | Distance and the Pythagorean Theorem | 39D | Students apply the Pythagorean Theorem to a real-world context. |
| 40 | Univariate Categorical Data | 40A | Students consider variations of categorical data. |
| 40 | Univariate Categorical Data | 40B | Students identify examples of categorical data. |
| 40 | Univariate Categorical Data | 40C | Students create a display for categorical data. |
| 40 | Univariate Categorical Data | 40D | Students convert a bar graph to a pie graph. |
| 41 | Univariate Quantitative Data | 41A | Students create data for a mean. |

| Big Idea No. | Big Idea | Task | Description |
|---|---|---|---|
| 41 | Univariate Quantitative Data | 41B | Students interpret a dot plot. |
| 41 | Univariate Quantitative Data | 41C | Students interpret a box plot. |
| 41 | Univariate Quantitative Data | 41D | Students create a display for a unique mean and median. |
| 42 | Displays of Univariate Quantitative Data | 42A | Students sketch a box plot related to given data. |
| 42 | Displays of Univariate Quantitative Data | 42B | Students interpret a dot plot. |
| 42 | Displays of Univariate Quantitative Data | 42C | Students create a dot plot and a box plot for given data. |
| 42 | Displays of Univariate Quantitative Data | 42D | Students draw conclusions from two box plots. |
| 43 | Deviations From the Mean | 43A | Students consider deviation from the mean with dot plots. |
| 43 | Deviations From the Mean | 43B | Students find absolute deviation. |
| 43 | Deviations From the Mean | 43C | Students create a dot plot for known mean and deviation. |
| 43 | Deviations From the Mean | 43D | Students interpret data to create a new dot plot. |
| 44 | Bivariate Categorical Data | 44A | Students consider and interpret bivariate data. |
| 44 | Bivariate Categorical Data | 44B | Students consider and interpret bivariate data. |
| 44 | Bivariate Categorical Data | 44C | Students consider and interpret bivariate data. |
| 44 | Bivariate Categorical Data | 44D | Students draw conclusions based on bivariate data. |
| 45 | Bivariate Quantitative Data | 45A | Students create a scatter plot and the situation it describes. |
| 45 | Bivariate Quantitative Data | 45B | Students use a scatter plot to describe association. |
| 45 | Bivariate Quantitative Data | 45C | Students draw conclusions using a scatter plot. |
| 45 | Bivariate Quantitative Data | 45D | Students compare different scatter plots to consider strong correlation. |

# ACKNOWLEDGMENTS

*Mine the Gap* has been a collaborative effort. We are grateful to Corwin for making this project a reality. We are thankful that they recognize the importance of quality mathematics tasks, anticipating student thinking, and considering what we do next. We appreciate that they, too, know that student answers aren't random, that flawed answers provide great insight, and that seemingly correct answers don't always tell the whole story.

We would like to especially thank the following teachers and coaches who embody the best of mathematics education. They were more than happy to try new tasks and gather examples of student thinking. They inspire us daily. Those colleagues are Laura Behrens, Graig Brubaker, Gail Bruton, Kaitlyn Carbaugh, Christy Dellota, Jenna DeMario, Christine Donnelly, Claudia Eckstrom, Megan Gittermann, Danika Gribble, Lindsay Kelley, Jennifer Leasure, Jennifer Lloyd, Kristen Mangus, Jennifer Marker, Alison Mariano, Jessica Michaelson, Stephanie Neagle, Nadege Nyemenim, Savana Pitney, Colleen Pollitt, Greta Richard, Jennifer Roberts, Kristen SanGiovanni, Molly Schaefer, Samantha Simpson, Rebecca Tees, AnnMarie Varlotta, Melissa Waggoner, Gavin West, Alexandra Weyforth, and Elizabeth Zinger.

Thanks to the staff at Corwin for transforming a featureless document into such an appealing, practical tool for teaching and learning. Special thanks to Erin Null for her enthusiasm, partnership, thoughtful questions, and insight.

Lastly and most importantly, many thanks to our families, who saw us less than they might like and tolerated us when we grumbled. We cannot thank them enough.

# ABOUT THE AUTHORS

 **John SanGiovanni** is a mathematics supervisor in Howard County, Maryland. There he leads mathematics curriculum development, digital learning, assessment, and professional development for 41 elementary schools and more than 1,500 teachers. John is an adjunct professor and coordinator of the Elementary Mathematics Instructional Leader graduate program at McDaniel College. He is an author and national mathematics curriculum and professional learning consultant. John is a frequent speaker at national conferences and institutes. He is also active in state and national professional organizations and currently serves on the Board of Directors for the National Council of Teachers of Mathematics.

 **Jennifer Rose Novak** is a mathematics education associate for the Delaware Department of Education. She provides leadership in the implementation of standards-based curriculum, instruction, and professional development in mathematics education to the 19 school districts and the various charter schools in Delaware. Jenny has served as a supervisor and coach in Howard County, Maryland, as well as an adjunct professor at McDaniel College and UMBC. She serves on a number of organizational boards, including MATHCOUNTS, the Delaware Mathematics Coalition, and the Maryland Council of Teachers of Mathematics, for which she is currently serving as president.

# INTRODUCTION

## MINING THE GAP

The pursuit of natural resources and precious gems is complex and hazardous. Yet, countless men and women pursue the work because the reward is so great. These individuals rely on investigation and research to determine when and where to strike. They make use of practical tools and innovation. They anticipate what will happen and plan accordingly. They are careful of unstable footing and faulty supports or bracing.

In some ways, we might think of teaching and learning mathematics similarly. Our students' success is precious to us. Achieving it is complex, and some days it even feels hazardous. We too make use of investigation and research to identify when and what to teach. Gaps in student understanding can create considerable consequences. Mathematics tasks are the tools that enable us to uncover those gaps. And high-quality mathematics tasks provide even greater insight into the quality, depth, and complexion of our students' understanding. Misconceptions and seemingly correct answers can undermine our stable footing and consistent progress.

Essentially, we must mine the gaps in our students' understanding so that we can achieve the goals we set for them. We must select and implement high-quality mathematics tasks to uncover student reasoning. We must anticipate what might happen, why it might happen, and what we will do about it when it happens.

## ABOUT THIS BOOK

Our mathematics instruction must be vibrant and engaging. We must go beyond direct instruction of rules and procedures. We must make use of much more than abundant practice and tedious worksheets. We can realize this vision with high-quality tasks that promote reasoning and representation.

Students have their own ideas about mathematics. Like their fingerprints, their ideas are unique to them. Sometimes, their reasoning is sophisticated, and other times, their reasoning is faulty. Their mistakes are not random. Instead, their mistakes are grounded in incomplete understanding, skewed observations, or flawed logic. We must uncover their confusion so that we can make informed decisions about our instructional next steps, and high-quality tasks enable us to do so.

This book is a collection of high-quality tasks aligned to the big ideas of intermediate mathematics. It shares perspective about what we might anticipate before our students work with specific tasks. It uncovers misconceptions, incomplete understanding, and unique student perspectives through multiple student work samples for each task. It offers what we might do next for the student samples. Three additional tasks are provided for each topic. The result is 180 examples of student work and 180 quality tasks for classroom use.

All of the tasks are provided electronically for use in the classroom. Options for modifying the tasks are provided so that any task can be used in any intermediate classroom.

## WHO IS THIS BOOK FOR?

This book is designed for any stakeholder. Classroom teachers can use the tasks in this book for instruction or assessment. Student work samples can be reviewed to better understand what might happen in the classroom as well as what might be done next.

Mathematics coaches and curriculum specialists can use this book to support the design of instruction or assessment resources. It can serve as a core resource to develop professional learning around characteristics of high-quality tasks, anticipating student thinking, identifying student misconceptions, and planning instructional next steps. This book can be a centerpiece for a school-wide professional learning community, an after-school workshop series, or collaborative planning meetings.

Principals and school administrators may also use the book to guide collaborative reviews of student work by grade-level teams. Administrators may rely on it to guide the development of common assessments for benchmarking student growth. Teacher preparation programs might use it for secondary methods or diagnosis and intervention courses.

## WHAT IS THE PURPOSE? WHAT IS THE PROBLEM?

Teaching middle school mathematics is a complicated endeavor. As teachers, we must understand the concepts, procedures, and application of seemingly "basic" mathematics. We must understand and apply research-informed pedagogy. We must also understand mathematical misconceptions and apparently correct answers as well as the "logic" behind these answers.

Misconceptions and incomplete mathematical thinking may go unnoticed because we are trained to think about mathematics in black and white or correct and incorrect. As teachers, we find incorrect answers and often react with steps to reteach the correct processes. However, student misconceptions are not random. Often, we may overlook why or how the answer was generated, and so the underlying problem persists. To complicate matters, even our students' correct answers do not always indicate accurate reasoning. For example, a student might correctly compare $\frac{3}{4}$ and $\frac{1}{2}$.

Yet, that student might select $\frac{3}{4}$ because both digits are larger rather than reason about the meaning of fractions and comparison.

As teachers, our lack of training for diagnosis is only part of the challenge. Teacher access to high-quality tasks that provide rich insight into student thinking can be limited. It can be a challenge to identify these tasks and even more challenging to create them. Yet, such tasks are necessary because of another complication with diagnosing student thinking. Unlike the perception of black-and-white answers in mathematics, student thinking and reasoning are highly variant. Simply, mistakes and misunderstandings do not always occur in the same ways or for the same reasons. They can be simplistic or complex. They can be independent of or connected to other mathematical skills and concepts.

There are two long-term ramifications of limited insight into student understanding. First and foremost, unrecognized student misconceptions become unchanged misconceptions. These misconceptions can become permanent ways of thinking. Each new layer of mathematics knowledge is then built on flawed foundations. These students are then likely to develop other misconceptions and forced to rely on rules and procedures that are lost over time. The other ramification is that the pattern continues in this teacher's classroom year after year, affecting large numbers of students.

## HOW DOES THIS BOOK SOLVE A PROBLEM?

As teachers, we need easy access to practical, quality tasks that uncover student thinking. We need multiple examples of tasks to support instruction throughout the year. We also need help thinking about how tasks are selected, what students might do, and what to do with student responses.

Low-level tasks featuring simple recall, procedures, or algorithms are often found in textbooks. Often, these tasks yield little more than correct or incorrect answers. They provide limited insight into our students' thinking. These shortcomings can challenge us to make informed instructional decisions. Moreover, inaccurate perceptions of our students also lead to instructional missteps. Student progress and long-term retention of skills and concepts are affected.

Conversely, the rich tasks featured in this book can be used for instruction or formative assessment. They will provide opportunities for teachers to go deeper with student performance by

- considering what students do and don't know about the content,
- describing misconceptions and limited understanding through incorrect and correct responses,
- anticipating what might happen with a specific task, and
- identifying possible instructional next steps.

As a result of using these books, teachers will be able to

- identify and select rich tasks for instruction or assessment,
- consider what counts as mathematical understanding,
- anticipate and plan for student misconceptions,

- make instructional decisions based on specific misconceptions or incomplete understandings, and

- access a substantial collection of rich tasks for classroom use.

## ORGANIZATION OF THE BOOK

This book is organized around the big ideas of mathematics in Grades 6 through 8, including

- operations with fractions and decimals;

- operations with integers;

- reasoning with ratios, proportions, and percents;

- representing and manipulating expressions, equations, and functions;

- understanding key geometry concepts; and

- analyzing statistical data.

Each chapter provides

- a collection of tasks aligned to the subtopics of the big ideas,

- a brief description of each task and its importance,

- ideas about what we might anticipate our students will do with the task,

- samples of student work with descriptions of what they did with the task,

- considerations for next steps with the highlighted student work,

- three additional tasks aligned to the mathematics topic, and

- ideas about what students might do with these additional tasks.

## THE APPROACH TO STUDENT WORK

This book is a guide to rich tasks, student understanding, and misconceptions. It is shaped by authentic student work and reasoning. The student work samples are from real students in real mathematics classrooms. The tasks were collected from random classrooms so that we could see what students do when working with these tasks. Tasks were not continuously distributed until just the right samples could be found. The tasks were provided after concepts had been taught, although in many cases, it had been weeks since the concept had appeared in the classroom. None of the student samples were collected within the same period as the concept was taught. Specific student samples were selected because they represent what our students frequently do or think about the mathematics.

## COMPANION WEBSITE

All of the tasks are provided electronically on the book's companion website at **resources.corwin.com/minethegap/6-8**. Note that the display of some tasks has been modified to fit the book's layout, but the full and complete versions can be found online. They can be reproduced for instruction, assessment, independent practice, or possibly for homework. Electronic downloads enable the reader to modify the tasks. Suggestions for modifications are made throughout the book.

# Three Reasons You Need This Book

1.  This book provides a wealth of high-quality mathematics tasks.

    Identifying high-quality mathematics tasks can be quite difficult. We can search online for hours trying to find just the right combination of rigor, relevance, and interest. Even then, we may not find what is best for our students. Writing or creating these tasks is even more difficult. This book provides 180 tasks electronically that can be modified, enhanced, or replicated for countless possibilities.

2.  This book provides insight into common approaches and misconceptions students have.

    Anticipating and identifying what students might do and the misconceptions they have can be acquired with years of experience. Yet, experience alone may not provide enough insight into what might happen with the students we're teaching this year. This book provides diverse student samples for 45 tasks related to the big ideas of mathematics in Grades 6–8.

3.  This book offers ideas about what we might do next.

    We know that "louder and slower" is not the solution to incorrect student thinking. Instead, we have to consider what students know and how well they understand it. We need to pinpoint where the mathematics falls apart for our students and determine what to do next. This book highlights where things go right and where they go wrong for students. It also gives ideas about next steps for reteaching, enriching, or extending our students' thinking.

# CHAPTER 1

# RICH MATHEMATICS TASKS, STUDENT MISCONCEPTIONS, USING TASKS

## PROMPTS WITH PURPOSE: USING HIGH-QUALITY TASKS

The quality of our mathematics tasks directly affects the learning of our students (Stein & Lane, 1996). For years, there has been a belief that the quantity of mathematics trumps the quality of mathematics. And so, it was likely believed that a practice sheet with 20 or more problems was a more effective way to learn mathematics than one or two high-quality prompts or tasks.

This notion of quantity creates a myriad of problems. For one, a student who practices a skill or concept incorrectly over and over again ingrains a misconception that can be extremely difficult for us to correct. Copious amounts of low-level practice become mundane and can cause students to fall out of love with mathematics. Often, these low-level tasks don't further one's learning. They don't always, if ever, promote reasoning. The procedural focus of isolated concepts may limit our students' ability to transfer mathematical ideas to new situations. Most important, low-level tasks don't provide opportunities to engage in mathematics in the same ways we encounter it in day-to-day life. In other words, mathematics in the real world isn't scripted. We don't continue to do or use the same skill over and over in a short amount of time. In the real world, mathematics isn't isolated and certainly isn't contrived.

Selection of high-quality mathematics tasks is a foundational part of exemplary mathematics instruction. After selecting tasks, we plan for our students to work with partners or small groups. Then, we must anticipate what will happen. We must consider the questions we will ask. We must think about how we will facilitate meaningful discourse and close the lesson. All of this builds from and with conceptual understanding and mathematics vocabulary. Simply, quality mathematics tasks alone won't produce proficient students. However, proficiency can't be developed without quality tasks.

## Tools of the Trade: Qualities of High-Quality Tasks

We might think of mathematics tasks as tools of our trade. Like tools designed for other jobs, we want the highest quality. We look for precision, craftsmanship, effectiveness, and practicality. The idea of high-quality mathematics tasks means different things to different people. There are all sorts of tools and rubrics for identifying a quality mathematics task. These tools usually identify that high-quality mathematics tasks

- align to mathematics content standards and/or significant mathematical ideas;
- make use of representations;
- provide students with opportunities for communicating their reasoning;
- can be modified for multiple entry points;
- create opportunities for different strategies for finding solutions;
- allow students to make connections between concepts;
- require cognitive effort; and
- are problem-based, authentic, or interesting.

## Selecting High-Quality Tasks

We can find mathematics tasks in textbooks, supplemental resources, and, of course, online. But how do we know if they are high quality? A rating or review tool can help us develop our "high-quality filter" for selecting good tasks. We can apply it to those tasks we find in print resources and online. It is important to keep in mind that there is no perfect task. Every task can be improved. This tool aligns to the characteristics of high-quality tasks described here.

---

### Identifying High-Quality Tasks

The purpose of the task is to teach or assess:

| ☐ Conceptual understanding | ☐ Procedural skill and fluency | ☐ Application |
|---|---|---|

**Rating Scale:**

2 - Fully Meets the Characteristic

1 - Partially Meets the Characteristic

0 - Does Not Meet the Characteristic

| The mathematics task | Rating |
|---|---|
| Aligns to mathematics content standards I am teaching. | |
| Encourages my students to use representations. | |
| Provides my students with an opportunity for communicating their reasoning. | |
| Has multiple entry points. | |
| Allows for different strategies for finding solutions. | |
| Makes connections between mathematical concepts, between concepts and procedures, or between concepts, procedures, and application. | |
| Prompts cognitive effort. | |
| Is problem based, authentic, or interesting. | |

---

Visit this book's companion website at **resources.corwin.com/ minethegap/6-8** for a downloadable version of this chart.

## The Purpose of the Task

Mathematical rigor promotes instructional balance between concepts, procedures, and applications of mathematics. So our initial consideration for selecting a task is to determine the purpose of the task. How does it connect with one of these components of rigor? Does the task engage students in concepts? Does it build procedural fluency? Does it apply concepts and procedures to problems and real-world contexts? There are other important considerations to determine the quality of the task.

## The task aligns to the mathematics standards I am teaching.

Tasks must be worthwhile and aligned to the skills and concepts in our curriculum.

| Tasks that fully meet this characteristic align directly to standards in my curriculum. | Tasks that partially meet this characteristic align to a standard in an adjacent grade level but are important and necessary. | Tasks that do not meet this characteristic do not connect with standards in my curriculum. |
| --- | --- | --- |

## The task encourages my students to use representations.

Representations help students make sense of and communicate mathematical ideas.

| Tasks that fully meet this characteristic explicitly direct students to use representations. | Tasks that partially meet this characteristic imply or provide space for representations. | Tasks that do not meet this characteristic are clearly procedural with no reference or space for representations. |
| --- | --- | --- |

## The task provides my students with an opportunity for communicating their reasoning.

Students can communicate their reasoning with models or pictures, numbers, and words.

| Tasks that fully meet this characteristic explicitly direct students to communicate their reasoning. | Tasks that partially meet this characteristic imply that students should communicate their reasoning. | Tasks that do not meet this characteristic do not require students to explain or justify their thinking. |
| --- | --- | --- |

## The task has multiple entry points.

Students can approach a problem from various perspectives, using diverse strategies and/or representations.

| Tasks that fully meet this characteristic are open to many possible solution paths and representations. | Tasks that partially meet this characteristic can be approached in different ways but may provide an example or prompt to direct students to an approach. | Tasks that do not meet this characteristic have a specific solution path intended or directed. |
| --- | --- | --- |

## The task allows for different strategies for finding solutions.

Students can solve a problem in various ways.

| Tasks that fully meet this characteristic are open to any strategy regardless of the efficiency of the strategy. | Tasks that partially meet this characteristic can be approached in different ways but imply a specific strategy for students to use. | Tasks that do not meet this characteristic direct students to a specific solution path or calculation. |
| --- | --- | --- |

## The task makes connections between mathematical concepts.

Mathematics ideas are related. We can also connect them to representations, procedures, and applications.

| Tasks that fully meet this characteristic connect mathematical ideas or connect concepts/procedures/applications within a topic. | Tasks that partially meet this characteristic allow for connections but do not call for them directly. | Tasks that do not meet this characteristic make no connections. They focus on a single procedure or recall. |
| --- | --- | --- |

## The task prompts cognitive effort.

High-quality tasks should generate some amount of struggle. Students should have to make sense of the prompt, the problem, or the representation.

| Tasks that fully meet this characteristic offer no obvious solution path. Or, tasks require concepts and procedures to be applied to new situations or contexts. | Tasks that partially meet this characteristic are problem based but indicate how they can be solved. | Tasks that do not meet this characteristic provide no cognitive resistance. Students are directed to do something exact or recall a skill or concept. |
| --- | --- | --- |

## Tasks are problem based, authentic, or interesting.

High-quality tasks are problem based. They can reflect real-world, authentic applications of mathematics. They should have interesting or novel prompts that grab students' attention.

| Tasks that fully meet this characteristic are problem based and authentic or interesting. | Tasks that partially meet this characteristic are problem based. | Tasks that do not meet this characteristic are not problem based. |
| --- | --- | --- |

## Hazards of Low-Level Tasks

Low-level tasks are typically grounded in recall and might only require procedure for completion. They lack the characteristics that engage students in reasoning about mathematical ideas. They don't provide opportunity for discussion. These tasks present mathematics concepts in isolated fashions.

Low-level tasks create other challenges that are difficult for us to overcome. Most important, these tasks provide a limited, if any, window into student understanding because they generally require a single, correct answer. They also require a specific strategy or process for successful completion. Because of this, we aren't always able to see our students' strategies, partial understanding, or misconceptions. Without this understanding, we are left to make guesses about what to do next. We are unable to directly address specific mathematics needs, so we reteach or provide interventions that don't necessarily address the problem.

Low-level tasks may also yield correct answers with flawed or incomplete understanding of the mathematics. A student may find the product of 5 × 6 by skip-counting by 5. He may apply this strategy to every multiplication expression he encounters. It may be his only strategy. If we consistently use low-level, basic multiplication prompts, we may only recognize his correct answers without learning of his only strategy. In time, his limited strategy will create considerable challenges because of the difficulty of skip-counting efficiently and accurately with larger, multi-digit numbers.

Low-level tasks

- make use of simple recall or procedure,

- require one answer that is found with a specific strategy or pathway,

- do not feature opportunities for representing mathematics,

- do not prompt for reasoning and justification, and

- lack connections within and between mathematics concepts.

## Is It REALLY a Good Task?

Some tasks are quite easy to identify as low level; typically, they are pure computation. But some tasks can be quite misleading. We must keep in mind that a task that makes use of representations does not automatically qualify as a quality task. Consider prompts that ask students to identify a fractional piece of a rectangle or circle. Sure, this presents the concept with a representation, but it is simply recall of what a fraction is. In other situations, we may see a representation that is different from those *we* encountered as students. The following task is an example. For many of us, we did not use tape diagrams to represent ratios. This "new" representation may mislead us to believe that a task that uses this new representation is of high quality. In fact, this task is an example of simple recall.

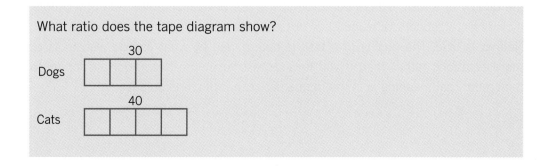

Another faulty indicator of quality is relying on tasks that require more than one right answer. Consider the following prompt.

Identify all of the integers below greater than −48.
- A. −19
- B. 37
- C. −49
- D. −60
- E. 25
- F. −100

The prompt above does ask for more than one right answer. However, the cognitive demand required to complete the task is quite low. We should look for tasks that have more than one strategy or solution path rather than more than one correct answer.

Context can also mislead us to the quality of a task. The purpose of learning mathematics is to apply it to the problems we face in the real world. So, it makes sense that high-quality tasks have a real-world connection. But when considering a task, we need to keep in mind if the real-world application makes sense. Is it contrived? Is it possible? Is the problem worth solving? Consider this problem:

32 ice skaters broke their arm at a figure skating competition. A hospital uses $5\frac{3}{8}$ feet of fabric to make a cast. The hospital has 300 yards of fabric available. Does the hospital have enough fabric for all of the skaters?

The task certainly provokes questions. Our students might ask, How many people were in the competition? Did their parents have to sign a waiver? Why does every cast get the same amount of fabric when each person's arm is different? Who measures out the fabric exactly like that? Why did they all go to the same hospital? Yet, none of these questions are about the mathematics. Why should they be? The problem is a silly fabrication so that fractions, measurement, and multi-step problem solving can be "applied" to an authentic situation. It reminds us of the classic "two trains leaving Chicago" problem.

Some believe that using manipulatives and tools is a "must-have" to determine the quality of a task. Others believe that the use of manipulatives and other tools, including calculators, cheapens or lessens the quality of a mathematics task. Neither is true. The quality of a task is determined by what you do with tools or representations rather than if you have access to them. It is important for us to remember that meaning is in the mathematics, not the manipulative.

We can use the quality tasks we select in different ways. We might choose to use them as our instructional centerpiece during a lesson. When doing this, we must consider if students will engage in the task independently or cooperatively. We must think about what our students might do and why they might do it. We also have to consider how we will debrief the task and facilitate discussion about the task, solution, and strategies for solutions.

We can also use these tasks for formative or summative assessment. When assessing with the tasks, we should reflect on what will satisfy the prompt. We should think about what ideas, representations, or strategies will count as evidence of understanding. We also want to begin to think about how we will use the information for our instructional next steps.

## Using Tasks Instructionally

Use of quality tasks for instruction, from selection to implementation, must be intentional. An effective way to implement these tasks is through three stages.

During the first stage, we set the context for the problem, revisit skills or concepts that students might use with the task, and convey our expectations for quality work and collaboration.

In the second stage, students engage collaboratively with partners or small groups to exchange ideas, apply strategies, and adjust thinking. During the second stage, we circulate to monitor student thinking. We ask questions to focus student thinking and make note of work that we want to highlight during the debrief. At this time, we might begin to think about the sequence of student sharing. We can give students numbers written on sticky notes to help organize our sequence. As we sequence their work, we want to consider how their relationships, representations, strategies, misconceptions, and errors are connected. We want to sequence so that each new sharing builds from or contrasts with the previous idea.

During the third stage, we facilitate a discussion or gallery walk so that groups can share their solutions and strategies. In this stage, students construct their meaning of the mathematics. They share their thinking and push back on the thinking of others. At this time, we facilitate a discussion, being careful to avoid influencing or even contaminating student thoughts by offering or dismissing specific strategies that we would use or prefer.

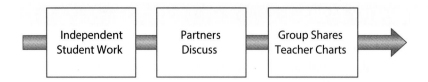

In some situations, we might modify the approach so that students engage with tasks independently before sharing ideas with partners and eventually the whole group. In this sequence, students work with the task independently, they share their ideas with a partner, and then the whole class comes together to share strategies and insights.

## Anticipating Student Responses

We must anticipate what our student responses may be so that our tasks, questions, and discussion are most effective. It is impossible to anticipate every strategy or misconception that our students will have. We can't expect to know every mistake they will make. Even so, anticipating possibilities prepares us for the instructional next steps we might take during the discussion, later in the lesson, or the next day. We can develop our ability to anticipate student responses by working with other teachers to select and plan tasks. We can work to complete the problem in as many different ways as possible, and we can even attempt to apply the misconceptions we think our students might have. Anticipating student thinking is highlighted throughout this book.

## Misconceptions

Anticipating student work enables us to imagine the misconceptions. Misconceptions are not random. They happen when students apply faulty logic. Often, misconceptions can be explained: They occur when students make improper connections between skills or concepts. Sometimes, this happens as students look to find patterns and connect them to seemingly similar procedures. For example, when adding fractions, we add numerators while the denominators remain unchanged. Our students might then do something similar when multiplying fractions by multiplying the numerators and keeping the denominators unchanged.

So what is a misconception? A misconception is any idea that is grounded in some degree of understanding but is mathematically flawed. Misconceptions can also be rules or strategies that work in certain situations. For example, we might compare $\frac{2}{8}$ and $\frac{5}{6}$, noting that $\frac{5}{6}$ is greater because we can think about the number of pieces missing from the whole. Yet, that strategy doesn't work when two fractions are missing the same number of pieces (e.g., $\frac{3}{4}$ and $\frac{9}{10}$).

Misconceptions can be developed through passive, independent observation and incomplete understanding. They can be taught by sharing chants and tricks or even simple "rules" or "have to's." We may unintentionally create them. For example, we may insinuate that there is always one solution for an equation when introducing the concept. But as we know, some equations have countless solutions when used to show the relationship between quantities. This contradiction can prove problematic for our students as they advance through mathematics classes.

If misconceptions go unnoticed, we are essentially reinforcing them. As teachers, we have to be on the lookout constantly for misconceptions. We have to probe thinking to be sure it is legitimate and complete. We have to be cautious of overreliance on correct answers. As we know, students can arrive at correct answers for the wrong reasons. Yet, correct answers can also be found with degrees of correct thinking.

Incomplete thinking can be just as hazardous as a misconception. *We might think of incomplete thinking as a mining hazard.* We assume that everything is fine because student answers and representations seem to indicate understanding. But in fact, our students have found a correct answer for the wrong reasons.

**MINING HAZARD**

**Look for this icon throughout the book. It highlights where students—and sometimes teachers—go awry.**

Consider a student who always works with region models of ratios. He consistently finds the shaded part and counts the total pieces. He may begin to associate the idea of a ratio as shaded parts to total parts. The shaded portion could also be considered in a part-to-part relationship. We also know that the unshaded portion can be expressed as a ratio in a part-to-part or part-to-whole relationship. He will give correct answers to shaded ratio problems. But he may not have a deeper understanding of ratio. Because of this, he may struggle tremendously with different representations, comparison of ratio, computation, problem solving, and unit rates.

Our preference for certain models may lead to challenges. For example, we might rely on integer chips and neglect number lines. But we may even mislead when we vary the models used during instruction. Consider working with integers on number lines. Students who consistently work with number lines that feature stagnant endpoints may develop incomplete understanding of integer relationships. This in turn contributes to challenges with computing integers and determining the reasonableness of answers.

In other situations, students may appear proficient because they show ability to apply algorithms to concepts. These students may appear to understand concepts like percent. Yet, they are unable to apply this "understanding" to certain problems, including those with percent-increase or percent-decrease contexts. They may give wildly inaccurate answers for situations with percents greater than a hundred because they have muddled the calculations. Or, they may rely on a pencil-and-paper calculation for problems like 50% of 86 or 25% of 12 that should be completed mentally.

### Facilitating Discourse

Discourse about our students' ideas and strategies is essential for maximizing the instructional potential of these tasks. These tasks can be applied seamlessly to the five practices Smith and Stein (2011) describe for orchestrating productive discussion. Those practices are

1. Anticipating student responses
2. Monitoring student work, engagement, and reasoning
3. Selecting student work for discussion
4. Sequencing student responses during discussion
5. Connecting responses and mathematical ideas

## Using High-Quality Tasks for Assessment

Quality assessments yield quality information about our students' understanding. These high-quality instructional tasks can easily be used for assessment purposes. We have to keep the purpose of assessment in mind as we use the task. Essentially, we have to know if we will use the task formatively or summatively. In other words, will we use it for instruction (formatively) or of instruction (summatively)?

### What Counts as Evidence?

Regardless of the assessment purpose, we must determine what will count as evidence of understanding before our students work with the task. We might ask if our students will have to compute. Will they be able to justify with a model or drawing? What might their drawings look like? What strategies might they use? Will they have to write sentences to convey understanding? All of these questions represent the thoughts we must consider when determining what counts as evidence of understanding.

This process is similar to anticipating what students will do with a task during instruction. Essentially, we want to determine what will constitute evidence of student understanding. We should make note of specific answers as well as the various strategies our students might use. It is important that we do not confine student responses to what we determine to be evidence.

### Understanding Student Thinking and Inference

We can be tempted to infer what students mean when we review student performance on assessments or assessment tasks, but this can be hazardous. High-quality tasks don't provide opportunities for random answers to be correct. But, as noted, correct answers are not always the result of correct or complete mathematics understanding. A good rule of thumb is to consider the question, "What would I ask him [the student] if he was here right now?" In other words, if we need to ask the student something about his work, his response, or his calculation, it is probably not complete. This doesn't necessarily have to affect a student's score or grade. Instead, it can be an indication of teaching or reinforcing that we need to do with the student.

## Determining Student Performance

We might think of our students' performance results in two distinct ways. In one instance, our students demonstrate understanding of the skills and concepts. These students are ready for more opportunities to reinforce their understanding or to advance it to more complex situations. Extending and enriching are other ways to think about advancing. In the other instance, our students demonstrate the need for reteaching.

| Reinforce and/or Advance | | Reteach | |
|---|---|---|---|
| Student demonstrates full understanding of the concept. The solution is correct. Reasoning is provided through pictures, words, or numbers/equations. Justification is complete. Minor errors may be present but do not affect the response. | Student demonstrates understanding of the concept. The solution may be incorrect but can be attributed to a computational error rather than flawed logic. Reasoning is provided but may not be complete. | Student demonstrates flawed logic or misconception. The solution may be correct, but it is coincidental. | Student demonstrates no understanding of the concept. The solution is incorrect. There is no justification or reasoning. Numbers or terms are disconnected from the prompt or a restatement of the prompt. |

## A General Rubric

We know that assessment is much more than counting the number of correct answers on a page. We know that there are layers to every answer. Rubrics are useful because they delineate the layers of understanding and performance. We can design specific rubrics for every task we use. Doing so can be quite daunting. Instead, we may choose to make use of a general rubric that can be applied to most, if not all, tasks. *We can also connect these rubrics with class observation sheets that capture evidence of student understanding during performance-based or hands-on tasks.*

 **MINING TIP**

**We can snap photos with our phones or tablets as students work on tasks to accompany our observation data. This can be helpful for communicating with parents.**

Visit this book's companion website at **resources.corwin.com/minethegap/6-8** for downloadable versions of this chart.

| Student Performance Recording Sheet Use this with formative assessment tasks, classroom activities, and observations. | | | | | |
|---|---|---|---|---|---|
| **Student Name** | Date | Date | Date | Date | Date |
| | | | | | |
| | | | | | |
| | | | | | |
| | | | | | |
| | | | | | |
| | | | | | |
| | | | | | |
| | | | | | |
| | | | | | |
| | | | | | |
| | | | | | |
| | | | | | |
| | | | | | |
| | | | | | |
| | | | | | |
| | | | | | |
| | | | | | |
| | | | | | |

**Reinforce and Advance**

Student demonstrates full understanding of the concept. The solution is correct. Reasoning is provided through pictures, words, or numbers/equations. Justification is complete. Minor errors may be present but do not impact the response.

Student demonstrates understanding of the concept. The solution may be incorrect but this can be attributed to a computational error rather than flawed logic. Reasoning is provided but may not be complete.

Student demonstrates flawed logic or misconception. The solution may be correct but it is coincidental.

**Reteach**

Student demonstrates no understanding of the concept. The solution is incorrect. There is no justification or reasoning. Numbers or terms are disconnected from the prompt or a restatement of the prompt.

## Using More Than One Task

As we know, it is important to triangulate data points to get a clear picture of where a student is mathematically. This is also true when using high-quality mathematics tasks. These tasks will give better insight into what students know, their partial understanding, and the misconceptions they have. Even so, we should be careful to avoid relying on one task as evidence of student understanding.

## Other Assessment Tips: Erasing

Erasing is a common practice. We make a mistake and we erase. It may be unavoidable. Yet, we may want to rethink why we erase and if it is always a good idea. When our students erase, they remove a strategy, diagram, or calculation that can provide good insight into their thinking. Knowing what students erase can be just as important as what they didn't erase. Granted, limited space and other confines may affect when our students have to erase. That being said, we may reconsider encouraging students to erase. Instead, we can include their errors as part of the written record.

## Other Assessment Tips: Hands-On Tasks

*This book provides a collection of paper-and-pencil tasks for instruction or assessment.* This is a limitation of the format and medium rather than a message of what we should value in a classroom or what makes a high-quality mathematics

**MINING TIP**

Other tips about the mathematics content, misconceptions, or implementing the tasks are sprinkled throughout the book. This icon signals those ideas.

task. It is critical that our students have hands-on, concrete experiences with tools, manipulatives, and other resources to learn mathematics. A hands-on task might be an opportunity when students are asked to compare two different fractions using two different tools such as fraction tiles and fraction circles. It's also important to keep in mind that the quality of hands-on tasks can vary wildly as well. The high-quality task identification tool presented earlier in this chapter can be used with hands-on experiences as well.

Hands-on tasks can also be applied to assessment situations. We can use observation rubrics, take notes of student statements and performance, and take pictures of what they do during the activity to document student understanding.

## Other Assessment Tips: Considering Bias

We must consider the representations and strategies that we prefer to use to model concepts or solve problems. This is important because it helps us uncover our natural bias toward certain methods even though our strategies may not be the only or best way to show or solve a problem. Acknowledging our preferences about mathematics concepts and problems can help us increase our ability to spot differing approaches and their validity.

We should also keep in mind that other experiences in mathematics class can influence our perspective of what a student does or doesn't understand. This might happen when a certain student is part of a group that shows understanding of a concept, but the individual student may not. It might happen when a student answers a question in class, but the solution is the result of coincidence rather than understanding. Student behavior may even influence our perspective of student understanding. Because of this, it may be wise to have students write their names on the back of their work so that we might see the evidence more clearly.

## Other Notes About Tasks: Modifying Tasks

Miners modify or customize their tools as the mining conditions change. *Each of the provided tasks can be modified.* We might do this so that they better align with our grade-level standards or for a different perspective on student thinking. Each task is provided electronically, and ideas for modification are sprinkled throughout.

**MODIFYING THE TASK**

**Look for this icon throughout the book. It calls attention to ideas for modifying the associated task.**

## Other Notes About Tasks: Content Connections

Mathematics concepts do not live in isolation. It is essential that we make explicit connections to other mathematics concepts within our grade and within the progression of ideas between grades. *There are content connections highlighted in this book that note how a task for a specific big idea builds on previous understanding or builds a foundation for later work.* When using these tasks for assessment, it's important to determine if errors or misconceptions are related to the focused idea or a connected idea so that appropriate next steps are taken.

**CONTENT CONNECTION**

- How do I typically find and select the mathematics tasks I use in my classroom?

- What are the characteristics of quality mathematics tasks?

- How do I think about the misconceptions my students have? Do I think about them before, during, and after a lesson?

- What misconceptions do I frequently encounter?

- How do I structure my mathematics tasks? Are they teacher centered or student centered? Are there opportunities for collaboration and discussion on a daily basis?

- What are my mining hazards? What are the skills, concepts, or topics that I teach that students seem to understand but do not fully understand?

- What experiences have I had modifying tasks when the topic is related but not completely aligned to what I am teaching?

# CHAPTER 2

# NUMBER SYSTEMS

**THIS CHAPTER HIGHLIGHTS HIGH-QUALITY TASKS FOR THE FOLLOWING:**

- Big Idea 1: Addition and Subtraction of Fractions

  Fractions are numbers that can be represented on number lines. We can make use of this model to add and subtract fractions as we do with whole numbers.

- Big Idea 2: Multiplication and Division of Fractions

  Different representations, including visual models and number lines, help to build conceptual understanding of multiplication and division of fractions.

- Big Idea 3: Reasoning About Addition and Subtraction of Fractions

  Reasoning about sums and differences of fractions provides students with opportunities to connect mathematical understanding to real-world situations and develop reasonableness of answers.

- Big Idea 4: Reasoning About Multiplication and Division of Fractions

  Reasoning about products and quotients of fractions provides students with opportunities to connect mathematical understanding to real-world situations and develop reasonableness of answers.

- Big Idea 5: Problem Solving With Fractions

  Problem solving with fractions provides students opportunities to model, reason, and justify strategies for generating solutions to problems with fractions.

- Big Idea 6: Decimals as Numbers

  Decimals are numbers that can be modeled in various ways, including number lines and number charts. Using these models helps students connect ideas about place value and decomposition of whole numbers to decimals.

- Big Idea 7: Addition and Subtraction With Decimals

  Different representations, including visual models and number lines, help to build conceptual understanding of addition and subtraction of decimals. Reasoning about operations with decimals helps us determine the reasonableness of our calculations.

- Big Idea 8: Multiplication and Division With Decimals

  Different representations, including visual models and number lines, help to build conceptual understanding of multiplication and division of decimals. Reasoning about operations with decimals helps us determine the reasonableness of our calculations.

# BIG IDEA 1
## Addition and Subtraction of Fractions

### TASK 1A

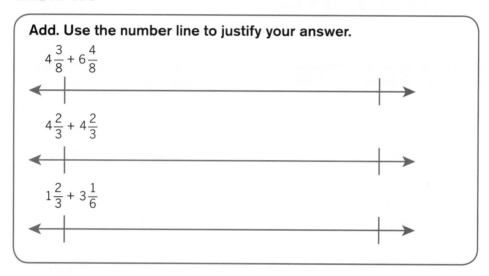

**Add. Use the number line to justify your answer.**

$4\frac{3}{8} + 6\frac{4}{8}$

$4\frac{2}{3} + 4\frac{2}{3}$

$1\frac{2}{3} + 3\frac{1}{6}$

## About the Task

Students grow their understanding of computation with whole numbers by making use of various models, especially number lines. Often, we make use of other models such as fraction circles and pattern blocks. In this task, students represent their understanding of adding mixed numbers on number lines. The number lines are open so that students are free to demonstrate understanding in various ways. Adding unlike denominators on a number line can be quite challenging unless we are careful to select favorable combinations of denominators including halves and fourths, fourths and eighths, thirds and sixths, or fifths and tenths.

## PAUSE AND REFLECT

- How does this task compare to tasks I've used?
- What might my students do in this task?

Visit this book's companion website at **resources.corwin.com/minethegap/6-8** for complete, downloadable versions of all tasks.

## Anticipating Student Responses

The first prompt requires students to add like denominators. Students may create endpoints that fully encompass the computation. These students will need to partition the intervals between whole numbers accordingly. Other students might begin with $4\frac{3}{8}$ as an endpoint on the number line and show a jump of 6 resulting in $10\frac{3}{8}$. From there, students may then show the additional jump of $\frac{4}{8}$ with a final sum of $10\frac{7}{8}$. Other students may use whole number tick marks and establish the position of $4\frac{3}{8}$ before adding on. These students may lose precision on their number line. The second prompt features like denominators that require regrouping. The strategies for adding these two addends are likely similar to those noted above. The third prompt will be more problematic for students because it makes use of unlike denominators and regroups a whole. Students will have to establish both thirds and sixths on the number line.

NOTES

### Student 1

Student 1 finds the sum of each prompt in this task accurately. Yet, his representations on the number line do not represent his work. He uses the first addend as a starting point for each number line. He then uses the whole number of the second addend for the number of jumps. But the size of his jumps are unit fractions related to the fraction in the second addend. The final location after adding on the number line does not match the sum of the equation above.

### Student 2

**MINING HAZARD**

**Right answers don't necessarily indicate correct thinking. Student 1 and Student 2 find correct sums but have limited or no conceptual understanding. It may be that the number line is an unfamiliar representation. We can investigate with other representations to confirm understanding before coming back to this model.**

*Like Student 1, Student 2 correctly finds the sums of each expression.* It appears that he can add the first two problems mentally as well as restate $8\frac{4}{3}$ as $9\frac{1}{3}$ mentally. He shows that he can add mixed numbers with unlike denominators in the third prompt with a common procedure. His first number line suggests that he can represent mixed numbers on a number line. He partitions by whole numbers and places his fractions relatively in between whole numbers. In his second prompt, he shows $4\frac{2}{3}$ at two different locations and incorrectly places $8\frac{4}{3}$ on the number line.

## USING EVIDENCE

*What would we want to ask these students? What might we do next?*

### Student 1

Student 1's work with this task is cause for concern. It appears that he can find the sum of mixed numbers. He appears to add the first two expressions mentally. There is evidence that he can use a procedure to find the sum in more challenging situations. One might argue that his mental calculations in the first two prompts challenge a need for this understanding. Yet, this is no clear indication that he is making sense of the addends or results. Working with the number line or other models can help develop his ability to determine the reasonableness of answers. We should revisit adding fractions less than 1 on a number line using common denominators and unlike denominators. We may introduce the strategy of beginning with one addend and counting on rather than beginning with 0 and making an initial jump of the first addend.

### Student 2

Student 2's first number line is encouraging. It shows some perspective about mixed numbers and number lines. We may ask him how it shows addition of the two mixed numbers. We may redirect him to show addition of whole numbers on a number line and ask him to compare that model with the first number line. This is unlikely to fully correct his thinking, but it does address the inconsistent representation. Like Student 1, we can then begin to reteach addition of fractions on number lines before moving to mixed numbers. There is a progression inherent in the reteaching.

# TASK 1A: Add. Use a number line to justify your answer.

$4\frac{3}{8} + 6\frac{4}{8}$ $\qquad$ $4\frac{2}{3} + 4\frac{2}{3}$ $\qquad$ $1\frac{2}{3} + 3\frac{1}{6}$

## Student Work 1

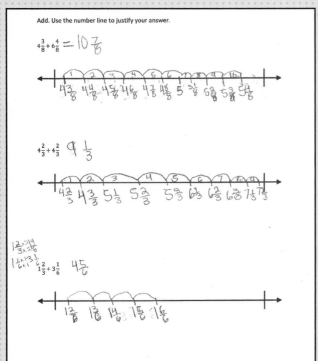

## Student Work 2

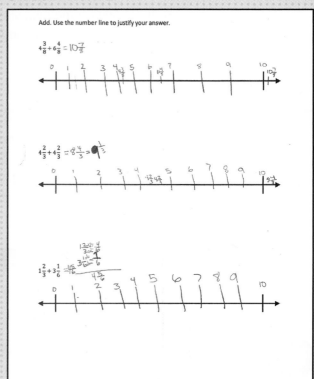

### Student 3

Student 3 finds the correct sum for each prompt. He breaks apart each addend and finds partial sums before recomposing. For $4\frac{3}{8}+6\frac{4}{8}$, he adds $\frac{3}{8}$ and $\frac{4}{8}$ on the number line, finding a sum of $\frac{7}{8}$. He then adds the whole numbers before finding a combined sum. He does something similar in the second prompt, even creating a new whole. He finds the correct sum in the third prompt but relies on a common denominator procedure for his solution.

### Student 4

Student 4's work is similar to Student 3 but not as complete. He justifies the addition of fractional parts or the mixed numbers in each prompt. He shows jumps on each number line but would be more accurate to note them as jumps of a specific fraction (e.g., $\frac{4}{8}$ instead of 4 in the first prompt). He finds common denominators in the third prompt to establish his jumps.

## USING EVIDENCE

*What would we want to ask these students? What might we do next?*

### Student 3

**MINING TIP**

Student 3's use of the number line was not the expected application. However, it is viable. It also provides an entry point for introducing partial sums with mixed numbers to students in the class.

*Student 3's work provides different insight into strategies for adding mixed numbers.* He makes use of partial sums. He uses the number line to represent how he adds the fractions. It is not surprising that the third representation is flawed as it is considerably more difficult. We can build on his understanding by adding whole number jumps to the number line. What's most important is that we recognize his viable strategy and encourage him to apply it to other situations to confirm that it always works.

### Student 4

Student 4's work represents adding fractions on a number line. He shows incremental jumps without established endpoints. This is reasonable. However, he doesn't account for the whole number addends. He would benefit from comparing his ideas to those of Student 3 to note how they are the same and different. In doing so, we want to highlight jumps of fractions instead of whole number jumps. We also want to focus on the whole number addends and discuss how they can be represented on the number lines as well.

$4\frac{3}{8} + 6\frac{4}{8}$  $\qquad$ $4\frac{2}{3} + 4\frac{2}{3}$ $\qquad$ $1\frac{2}{3} + 3\frac{1}{6}$

## Student Work 3

Add. Use the number line to justify your answer.

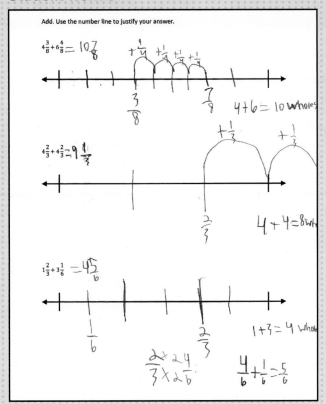

$4\frac{3}{8} + 6\frac{4}{8} = 10\frac{7}{8}$

$4 + 6 = 10$ wholes

$4\frac{2}{3} + 4\frac{2}{3} = 9\frac{1}{3}$

$4 + 4 = 8$ wh

$1\frac{2}{3} + 3\frac{1}{6} = 4\frac{5}{6}$

$1 + 3 = 4$ whole

$\frac{2 \times 2}{3 \times 2}\frac{4}{6}$  $\frac{4}{6} + \frac{1}{6} = \frac{5}{6}$

## Student Work 4

Add. Use the number line to justify your answer.

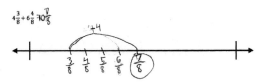

$4\frac{3}{8} + 6\frac{4}{8} = 10\frac{7}{8}$

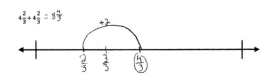

$4\frac{2}{3} + 4\frac{2}{3} = 8\frac{4}{3}$

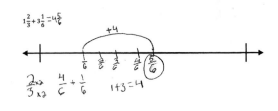

$1\frac{2}{3} + 3\frac{1}{6} = 4\frac{5}{6}$

$\frac{2 \times 2}{3 \times 2}$  $\frac{4}{6} + \frac{1}{6}$  $1 + 3 = 4$

## OTHER TASKS

- What will count as evidence of understanding?
- What misconceptions might you find?
- What will you do or how will you respond?

Visit this book's companion website at **resources.corwin.com/minethegap/6-8** for complete, downloadable versions of all tasks.

**TASK 1B:** Two fractions have a sum of $\frac{10}{16}$. What might those fractions be? Use a number line to explain your thinking.

Two fractions have a sum of $\frac{5}{8}$. What might those fractions be? Use a number line to explain your thinking.

**CONTENT CONNECTION**

As students develop fluency working with integers and negative quantities, this task may be introduced to see how students can show sums with positive and negative rational numbers.

*This open-ended task has many possibilities.* The first number line is partitioned into sixteenths and simply sets the stage for the second prompt. The first prompt helps us determine if students have a basic understanding of addition on a number line. The second number line and writing prompt allow for insight into student thinking about equivalent fractions. Some students will need to establish eighths by counting and skipping tick marks while others will simply note the connection with the prompt above. Yet others may count each tick mark as an eighth, creating an endpoint of 2 rather than 1.

**TASK 1C:** Robert subtracted $\frac{7}{12} - \frac{1}{2}$ and got a difference of $\frac{6}{10}$. Use what you know about fractions to explain why you agree or disagree with Robert. Use models, numbers, or words to explain your thinking.

This is a take on the classic misconception that students add denominators and numerators when computing with fractions. The task is primed to provoke this misunderstanding if students have it. We must be sure that students who hold this misunderstanding only apply it to subtraction. We must also determine if this misunderstanding occurs with both like and unlike denominator situations. It may be necessary to completely reteach and revisit addition and subtraction with fractions. In doing so, making use of fractions with uncomplicated denominators (thirds, fourths, sixths, and eighths) and in context may be most helpful.

**TASK 1D:** There were 12 pizzas for a team party. $8\frac{3}{8}$ of the pizzas were eaten by players. $2\frac{1}{2}$ pizzas were eaten by parents. The rest of the pizzas were taken home. How much pizza was eaten at the party? How much pizza was taken home? Use models, numbers, or words to explain your thinking.

This two-step problem can be approached in various ways. Students might take away the amount of pizza eaten by players and then the amount eaten by parents. Others might add the two amounts eaten and then take that sum away from 12. Students can represent their thinking with pictures, bar diagrams, or number lines. Some students will use equations to find their solutions. Regardless of strategy, we want to be sure that our students can explain and connect their representation to the problem. It would be wise to bring various representations together to help students find connections between strategies and develop their understanding.

NOTES

NOTES
_____

# BIG IDEA 2
## Multiplication and Division of Fractions

### TASK 2A

> Cam says that when he multiplies mixed numbers, he breaks them apart like whole numbers.
>
> He uses $4\frac{2}{5} \times 5$ as an example.
>
> He says $4\frac{2}{5} \times 5$ is the same as $4 \times 5 + \frac{2}{5} \times 5$.
>
> Use pictures, numbers, or words to explain why you agree or disagree with Cam.
>
> Give another example to justify why you agree or disagree with Cam.

## About the Task

For some reason, we do not always think of fractions in the same ways that we do whole numbers or even decimals. It makes perfect sense for us to break apart multi-digit numbers before multiplying them. We can think about mixed numbers in the same way. Essentially, we are applying the distributive property. In this task, students are asked to consider if decomposition of mixed numbers can be applied to multiplication of fractions.

## Anticipating Student Responses

Some students will disagree. They will argue that $4\frac{2}{5}$ is not the same as $4+\frac{2}{5}$. Others will make calculation errors that support their disagreement. Students who agree will also offer varied reasons. Some will procedurally find the product of both situations. Others will do so with a model or drawing. Students may also prove that $4\frac{2}{5}$ is the same as $4+\frac{2}{5}$, so if you multiply any number by the two values you will always get the same product. Some students may extend this thought to other operations or different mixed numbers.

### PAUSE AND REFLECT

- How does this task compare to tasks I've used?
- What might my students do in this task?

 Visit this book's companion website at **resources.corwin.com/minethegap/6-8** for complete, downloadable versions of all tasks.

### Student 1

Student 1 disagrees that the two expressions are equivalent. She computes to justify her conclusion. She compares the two products, noting that they "don't match" (are not equal). We can see why she has a faulty conclusion. She multiplies $4\frac{2}{5} \times 5$ by converting to $\frac{22}{5}$ and multiplying. When she multiplies the decomposed expression, she multiplies $4 \times 5$ and then *adds* $\frac{2}{5}$.

### Student 2

Student 2's rationale is rooted in the fact that $22 \neq \frac{110}{6}$. While her statement is correct in isolation, it is not correct relative to the mathematics of the problem. We find her problem with the decomposed computation. She begins by multiplying $4 \times 5$. She restates the product of 20 as $\frac{20}{1}$. She then adds $\frac{20}{1}$ and $\frac{2}{5}$, finding a sum of $\frac{22}{6}$. She multiplies $\frac{22}{6}$ and $\frac{5}{1}$,

finding a product of $\frac{110}{6}$, and notes that it is not the same as 22.

### Student 3

Student 3 agrees that the two expressions are equivalent because they both equal 22. She shows how to convert the mixed number prior to multiplying in the traditional manner. She establishes that the partial approach also results in the same product.

### Student 4

Like others, Student 4 uses computation to validate her solution. However, she makes a significant statement that could stand alone as evidence. She states that the two expressions [are] "just breaking it apart into two pieces and then adding them back together."

*What would we want to ask these students? What might we do next?*

### Student 1

We might review Student 1's work with her. We can ask her why she added $20 + \frac{2}{5}$. We can choose to revisit multiplying mixed numbers with models or review the work of others. Her work would be a good selection for the class to discuss. This conversation should help her understand why we have to multiply both 4 and $\frac{2}{5}$ by 5.

### Student 2

Student 2 has the right ideas about multiplying mixed numbers. A calculation error creates her problem. It may be an oversight, or it may be that she has muddled ideas and operations. In other words, she is working with multiplication, which computes "across numerators and denominators." And so, she may have inadvertently applied this process to addition. Like Student 1, we want to ask her about why she adds both numerators and denominators.

### Student 3

Student 3 has the computation in place to prove her thinking. However, she hasn't fully connected it to the concept of the distributive property. She should be exposed to the idea and have practice applying it to new situations.

### Student 4

Student 4 has functional understanding of the distributive property. We may extend her understanding to multiplying two mixed numbers. We can make use of symbolic representations or area models used for multiplying two two-digit numbers. It is important to keep in mind that partial products of any kind, especially two mixed numbers, may become less efficient or even less practical.

**TASK 2A:** Cam says that when he multiplies mixed numbers, he breaks them apart like whole numbers. He uses $4\frac{2}{5} \times 5$ as an example. He says $4\frac{2}{5} \times 5$ is the same as $4 \times 5 + \frac{2}{5} \times 5$. Use pictures, numbers, or words to explain why you agree or disagree with Cam. Give another example to justify why you agree or disagree with Cam.

## Student Work 1

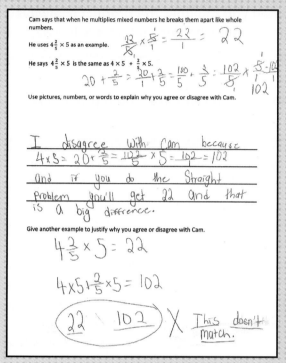

I disagree with Cam because $4 \times 5 = 20 + \frac{2}{5} = \frac{102}{5} \times 5 = \frac{102}{5} = 102$ and if you do the straight problem you'll get 22 and that is a big difrence.

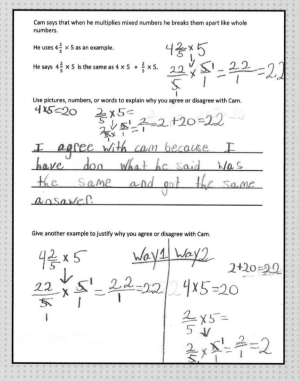

## Student Work 2

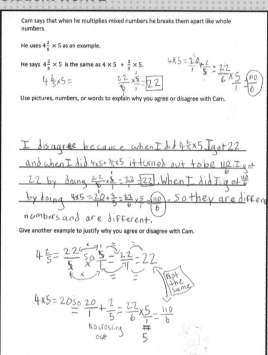

I disagree because when I did $4\frac{2}{5} \times 5$ I got 22 and when I did $4 \times 5 + \frac{2}{5} \times 5$ it turned out to be 110. I got 22 by doing $\frac{22}{5} \times \frac{5}{1} = 22$ [22]. When I did I got $\frac{110}{6}$ by doing $4 \times 5 = 20 + \frac{2}{5} = \frac{22}{6} \times 5$ ($\frac{110}{6}$). So they are differnt numbers and are different.

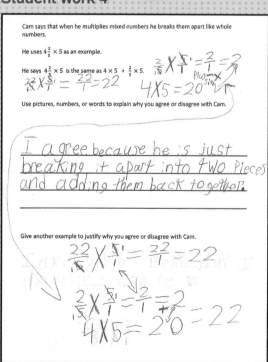

## Student Work 3

I agree with cam because I have don what he said was the same and got the same ansawer.

## Student Work 4

I agree because he is just breaking it apart into two pieces and adding them back together.

## OTHER TASKS

- What will count as evidence of understanding?
- What misconceptions might you find?
- What will you do or how will you respond?

 Visit this book's companion website at **resources.corwin.com/minethegap/6-8** for complete, downloadable versions of all tasks.

**TASK 2B:** Carson thinks that when you multiply, the product is always larger than both of the factors. For example, he knows that $4 \times 8 = 32$, and 32 is greater than 4 and 8.

Carson doesn't understand why $\frac{1}{3} \times \frac{3}{4} = \frac{1}{4}$.

Use models, numbers, or words to prove to Carson that the product is correct.

**MINING HAZARD**

Tasks that ask students to share conceptual understanding may not be best justified with procedural computations. Correct calculations may lead us to think that students understand the concepts.

*This task is designed to expose thinking about another misconception prevalent when multiplying and dividing.* The misconception is that multiplication creates a larger product and division creates a quotient less than the dividend. This isn't true with fractions. The computation is provided. Some students will use an array or area model to justify the solution. Other students may create an equal groups situation or communicate that it makes sense because the equation reads "one-third groups of three-fourths." Some students may note that one-third of three is one, so "one-third of three-fourths is one fourth." Look for students who multiply numerators and denominators finding a product of $\frac{3}{12}$, which is equivalent to $\frac{1}{4}$. Though the mathematics is correct, this doesn't necessarily communicate understanding.

**TASK 2C:** Andrea's dog eats $\frac{1}{4}$ of a pound of food each time he eats. Andrea has a 25-pound bag of food. How many times can she feed her dog with a full bag?

A painter has a 5-gallon bucket of paint. She can fill her paint tray with $\frac{3}{4}$ of a gallon. She will use all 5 gallons to paint a house. How many times will she refill her paint tray?

**MODIFYING THE TASK**

Change the fractions to align with the content being taught in your mathematics class. This might include dividing a mixed number by a fraction or a fraction by a whole number.

Making sense of quotients when dividing with fractions may happen best when the mathematics is put into context. *In Task 2C, students are presented with two story problems that require division of fractions.* In both cases, they will divide a whole number by a fraction. In this task, look for students who reason about the relationship and represent the problem with a drawing or equation. The second problem provides a good opportunity for conversation. Students will need to consider what to do with the extra half of a gallon after filling the tray *six* times.

**TASK 2D:** Chris is sure that when you divide with fractions, you get smaller numbers. In class, his group divides 4 by $\frac{1}{3}$ and gets 12. Chris disagrees with his group. Do you agree with Chris or his group? Use models, numbers, or words to explain your thinking.

*This task highlights the classic misconception about division with fractions.* It may be surprising how many students hold the idea that division yields smaller numbers. *Interestingly, some students may be able to procedurally complete the computation but not be able to make sense of why the result is larger.* In both cases, we need to revisit the concept with unsuccessful students. It is extremely important to use models and drawings while asking explicit questions related to the division context, including "How many thirds are in 1?", "How many thirds are in 2?", or "Why does it make sense that there are more thirds in two wholes than in one whole?" Begin reteaching with unit fractions before moving to more complicated fractions. We also want to be sure that we explicitly connect the equations to the representations.

NOTES

# BIG IDEA 3

# Reasoning About Addition and Subtraction of Fractions

### TASK 3A

Tell if the DIFFERENCE of each will be more than $\frac{1}{2}$ or less than $\frac{1}{2}$.

$\frac{9}{8} - \frac{1}{6}$　　　$\frac{3}{4} - \frac{6}{12}$　　　$\frac{9}{10} - \frac{1}{5}$　　　$\frac{6}{10} - \frac{8}{16}$

Choose one problem from above. Explain how you know the difference is more than $\frac{1}{2}$ or less than $\frac{1}{2}$.

## About the Task

How do we reason about sums and differences? Occasionally, we consider the amount that the minuend represents. In other situations, we think about the size that we take away. In this task, students are asked to consider subtraction results. The second part of the task provides insight into their reasoning. Some students can develop this reasoning through discussion and symbolic representations. Others benefit from discussion that makes use of concrete models or visual representations.

## Anticipating Student Responses

Some of our students will find exact answers to determine if the difference is more or less than $\frac{1}{2}$. This is not the intention of the task. The reasoning in the third prompt is most easy to relate to common denominators. It makes sense that our students would make use of that strategy, but we would prefer that it be done through recall and relational understanding rather than computation. In the other prompts, we want students to reason about the minuend and subtrahend. The first prompt takes a small quantity away from a value greater than 1, or $\frac{9}{8}$. We know that this difference will be greater than $\frac{1}{2}$. In the second and fourth prompts, we take $\frac{1}{2}$ away from a value less than one whole. When this happens, our result is less than $\frac{1}{2}$.

### PAUSE AND REFLECT

• How does this task compare to tasks I've used?

• What might my students do in this task?

Visit this book's companion website at
**resources.corwin.com/minethegap/6-8**
for complete, downloadable versions of all tasks.

## Student 1

Student 1 correctly compares the difference in the first prompt. His other comparisons are incorrect. His writing implies that he might not know the mathematical meaning for the term "difference." It also reveals inaccurate perceptions about comparison ("$\frac{6}{12}$ is greater than $\frac{3}{4}$"). With these ideas in mind, it is unlikely that his correct response to the first prompt is representative of his understanding of fractions.

## Student 2

Student 2's responses are correct. His work captures all of the procedures he used to find his comparisons. His work is accurate. Is it efficient? Is it based on reasoning about the meaning of the fractions being computed? Does he reason about the size of these fractions as well as their relationship to $\frac{1}{2}$? His writing provides an interesting note about subtracting $\frac{6}{10} - \frac{8}{16}$ by thinking of $\frac{8}{16}$ as $\frac{5}{10}$.

## Student 3

Student 3's use of precise fractions and his writing signal procedurally based thinking. In fact, his writing talks about *how* he computed.

## Student 4

Student 4 offers the reasoning the task is intended to provoke. He accurately compares the differences to $\frac{1}{2}$. There is no evidence of computation. He relates the minuend and the subtrahend to the $\frac{1}{2}$ benchmark. We should note that his reasoning is slightly flawed.

*What would we want to ask these students? What might we do next?*

## Student 1

Student 1 presents different challenges. Does he have incorrect responses because of the inability to compare or the inability to subtract? Does he understand that we can work with fractions greater than 1? In these situations, we have to find ways to drill down to a root cause. We can make use of other, skill-based tasks to get a better sense of the student's understanding. In this case, we should verify that he understands the meaning of a fraction both greater and less than 1. We should then confirm that he can compare fractions with like denominators and then unlike denominators. From here, we can determine his ability to subtract fractions.

## Students 2 and 3

We can be confident that Students 2 and 3 can rely on calculations to find solutions. Our next move is to work to develop their reasoning about the fractions being computed. We should revisit benchmark fractions and how they can support computation. As with other reasoning activities, it is critical that we provide students with opportunities to share and listen to strategies and approaches. During conversations, it is important to highlight clever approaches such as Student 2's swapping of $\frac{8}{16}$ with $\frac{5}{10}$.

## Student 4

As noted, Student 4's approach for comparing the difference to $\frac{1}{2}$ is slightly flawed. He seems to consider the impact of removing half from a value less than 1 by noting that $\frac{6}{10}$ is greater than $\frac{1}{2}$ but should instead note that $\frac{6}{10}$ is less than 1. With this in mind, Student 4 still encourages us about the reasoning potential of our students. He, too, will benefit from conversations about strategies and logic when working with these sorts of tasks. We should praise his insight, but be sure to further the discussion by challenging his logic as noted here.

**TASK 3A:** Tell if the DIFFERENCE of each will be more than $\frac{1}{2}$ or less than $\frac{1}{2}$.

$$\frac{9}{8} - \frac{1}{6} \qquad \frac{3}{4} - \frac{6}{12} \qquad \frac{9}{10} - \frac{1}{5} \qquad \frac{6}{10} - \frac{8}{16}$$

Choose one problem. Explain how you know the difference is more than $\frac{1}{2}$ or less than $\frac{1}{2}$.

## Student Work 1

Tell if the DIFFERENCE will be more than $\frac{1}{2}$ or less than $\frac{1}{2}$.

| Subtraction problem | More than $\frac{1}{2}$ or less than $\frac{1}{2}$ |
| --- | --- |
| $\frac{9}{8} - \frac{1}{6}$ | More |
| $\frac{3}{4} - \frac{6}{12}$ | More |
| $\frac{9}{10} - \frac{1}{5}$ | less |
| $\frac{6}{10} - \frac{8}{16}$ | More |

Choose one problem from above. Explain how you know the difference is more than $\frac{1}{2}$ or less than $\frac{1}{2}$.

$\frac{9}{8} - \frac{1}{6}$

$\frac{3}{4} - \frac{6}{12}$

$\frac{9}{10} - \frac{1}{5}$

$\frac{6}{10} - \frac{8}{16}$

The difference between $\frac{3}{4}$ and $\frac{6}{12}$ is that $\frac{6}{12}$ is greater than $\frac{3}{4}$ because you cant add it.

## Student Work 2

Tell if the DIFFERENCE will be more than $\frac{1}{2}$ or less than $\frac{1}{2}$.

| Subtraction problem | More than $\frac{1}{2}$ or less than $\frac{1}{2}$ |
| --- | --- |
| $\frac{9}{8} - \frac{1}{6}$ | More |
| $\frac{3}{4} - \frac{6}{12}$ | Less |
| $\frac{9}{10} - \frac{1}{5}$ | More |
| $\frac{6}{10} - \frac{8}{16}$ | Less |

Choose one problem from above. Explain how you know the difference is more than $\frac{1}{2}$ or less than $\frac{1}{2}$.

$\frac{6}{10} - \frac{8}{16}$ is an easy one. $\frac{8}{16}$ is equal to $\frac{1}{2}$ witch can convert to $\frac{5}{10}$ so I can easily subtract $\frac{6}{10} - \frac{5}{10}$ and my final answer is $\frac{1}{10}$ witch is less than a half.

## Student Work 3

Tell if the DIFFERENCE will be more than $\frac{1}{2}$ or less than $\frac{1}{2}$.

| Subtraction problem | More than $\frac{1}{2}$ or less than $\frac{1}{2}$ |
| --- | --- |
| $\frac{9}{8} - \frac{1}{6}$ | $\frac{23}{24}$ more |
| $\frac{3}{4} - \frac{6}{12}$ | $\frac{3}{12}$ less |
| $\frac{9}{10} - \frac{1}{5}$ | $\frac{7}{10}$ more |
| $\frac{6}{10} - \frac{8}{16}$ | $\frac{6}{80}$ more |

Choose one problem from above. Explain how you know the difference is more than $\frac{1}{2}$ or less than $\frac{1}{2}$.

$\frac{9}{8} - \frac{1}{6}$ is more than $\frac{1}{2}$ because if you find the common denominator it will be 24. It will be 24 because $\frac{9 \times 3}{8 \times 3}$ and $\frac{1 \times 4}{6 \times 4}$ make $\frac{27}{24} - \frac{4}{24}$ Which eaquals $\frac{23}{24}$ and $\frac{23}{24}$ is closer to 1 whole.

## Student Work 4

Tell if the DIFFERENCE will be more than $\frac{1}{2}$ or less than $\frac{1}{2}$.

| Subtraction problem | More than $\frac{1}{2}$ or less than $\frac{1}{2}$ |
| --- | --- |
| $\frac{9}{8} - \frac{1}{6}$ | more |
| $\frac{3}{4} - \frac{6}{12}$ | -Less |
| $\frac{9}{10} - \frac{1}{5}$ | more |
| $\frac{6}{10} - \frac{8}{16}$ | Less |

Choose one problem from above. Explain how you know the difference is more than $\frac{1}{2}$ or less than $\frac{1}{2}$.

I Know $\frac{6}{10} - \frac{8}{16}$ is less because $\frac{6}{10} \ge \frac{1}{2}$ & $\frac{8}{16} = \frac{1}{2}$, so the difference is less.

## OTHER TASKS

- What will count as evidence of understanding?
- What misconceptions might you find?
- What will you do or how will you respond?

Visit this book's companion website at **resources.corwin.com/ minethegap/6-8** for complete, downloadable versions of all tasks.

**TASK 3B:** Which addition expression has the greater sum? Tell how you know which sum is greater.

$$\frac{1}{4} + \frac{1}{2} \text{ or } \frac{1}{2} + \frac{1}{6} \qquad\qquad \frac{3}{4} + \frac{3}{2} \text{ or } \frac{3}{6} + \frac{3}{9}$$

In this task, students compare sums of different fractions. Reasoning about both prompts likely relates the addends, the sums, or both to benchmarks. But there are other strategies. In the first prompt, both expressions add something to $\frac{1}{2}$ so the expression that adds more will yield a greater sum. In the second prompt, we might reason that the second sum is less than 1 while the first sum is clearly greater than 1 because one of the addends is greater than 1. In both prompts, students may show misunderstanding of how to add fractions, while others add correctly, though finding the results is much less efficient.

**TASK 3C:** Tell if the sum of each will be greater or less than 12. Choose one of the expressions and tell how you know it is greater or less than 12.

$$3\frac{10}{12} + 8\frac{5}{6} \qquad 6\frac{1}{4} + 5\frac{3}{8} \qquad 6\frac{1}{2} + 5\frac{1}{4} + \frac{1}{8} \qquad 4\frac{3}{4} + 4\frac{7}{8} + 3\frac{1}{8}$$

We can't limit our reasoning activities to fractions less than 1. In each prompt, students reason about the sum and compare it to 12. Students may decompose the mixed numbers to add wholes before considering the fractions. For example, in the first prompt we can add the whole numbers 3 and 8, finding a sum of 11. We're left with $11 + \frac{10}{12} + \frac{5}{6}$. Both fractions are much larger than $\frac{1}{2}$, so 11 plus those fractions will clearly be greater than 12. In the past, students who can procedurally add these mixed numbers may lead us to believe that they have a strong sense of fractions and computation when that may not have been the case.

**TASK 3D:** Henry says that he knows that the difference of $7\frac{7}{8} - 2\frac{1}{2}$ is greater than 5 without subtracting on a piece of paper. How might he know this?

Henry also says that he knows the difference of $16\frac{2}{3} - 5\frac{1}{6}$ is greater than 11 without subtracting on a piece of paper. How might he know this?

This task asks students to reason about the results of subtracting mixed numbers relative to whole numbers greater than 1. It also prompts them to avoid using a procedure or paper/pencil for finding the difference. This will prove to be a substantial challenge for some of our students. They may disregard the fractional part of each number, noting that $7\frac{7}{8} - 2\frac{1}{2}$ isn't greater than 5 because $7 - 2 = 5$. Others may mistake the fractional amount being taken away. Others will reason that $7\frac{7}{8} - 2 = 5\frac{7}{8}$. They then share that $\frac{1}{2}$ from $5\frac{7}{8}$ is more than 5 because $\frac{7}{8}$ is more than $\frac{1}{2}$.

NOTES

NOTES

# BIG IDEA 4
## Reasoning About Multiplication and Division of Fractions

### TASK 4A

> **Which expression has the greatest quotient? Use models, numbers, or words to explain your thinking.**
>
> $20 \div \frac{1}{2}$ $\qquad$ $20 \div \frac{1}{4}$ $\qquad$ $20 \div \frac{1}{5}$

### About the Task

Reasoning about computation is useful for making sense of answers. This may even be more valid when we divide with fractions and decimals. *In this task, students are asked to determine which expression creates the greatest quotient.* In all three expressions, the dividend is the same (20). In whole number situations, greater divisors yield smaller quotients. This is not true when we divide with fractions. As we know, in each expression we are asking how many halves, fourths, or fifths are in each.

### Anticipating Student Responses

Students may apply the incorrect process and multiply the whole number by the fraction. Students may misunderstand the meaning of the fraction. *For example, they may interpret $20 \div \frac{1}{5}$ as $20 \div 5$.* Successful students may draw a representation of each expression. Some students may do the actual computation and then compare quotients. Some students will reason about the size of the fractions, noting that there will be more "smaller fractions" in a whole number.

 **MODIFYING THE TASK**

We can change the expressions to divide with fractions other than unit fractions as students develop and refine their understanding of division with fractions.

 **MODIFYING THE TASK**

Drawing $20 \div \frac{1}{5}$ may be quite complicated. Considering modifying the dividend and/or fractions so that students who rely on images have a better entry point. For example, $4 \div \frac{1}{2}$, $4 \div \frac{1}{4}$, and $4 \div \frac{1}{8}$ might be better options.

### PAUSE AND REFLECT

- How does this task compare to tasks I've used?
- What might my students do in this task?

 Visit this book's companion website at **resources.corwin.com/minethegap/6-8** for complete, downloadable versions of all tasks.

### Student 1

Student 1 has a noticeable misconception. He knows that halves are larger than fifths and fourths. Because halves are larger, he reasons that there will be more of them in 20. This is true for multiplication but not for division.

### Student 2

Student 2 incorrectly states that $20 \div \frac{1}{2}$ creates the greatest quotient. His argument seems connected to his comfort with halving. He shows that 10 is half of 20. He doesn't understand that this is $20 \div 2$ rather than how many halves are in 20, or $20 \div \frac{1}{2}$. He applies the thought of division to the other equations relative to whole number division. It may be possible that he has confused the procedures for multiplying and dividing fractions.

## USING EVIDENCE

*What would we want to ask these students? What might we do next?*

### Student 1

Student 1 may misunderstand the results of multiplication and division. He reverses the relationship between the size of the divisor and its impact on the quotient. We can explore this reasoning with whole numbers. We can then move to fractions, especially unit fractions, after confirming his understanding about the results of multiplication and division. Concrete models and pictures should be accompanied by questions such as, "What do you notice about the number of pieces as the size of the pieces get smaller?" We should also consider working with smaller dividends to establish understanding.

### Student 2

We need to work with Student 2 to make sense of what the equation is asking. We need to reinforce that it is asking how many halves are in 20 rather than what is half of 20. This may be difficult to establish. It may be helpful to work with unit fractions that feature denominators that aren't factors of the whole number. In this case, we might use thirds because $20 \div 3$ doesn't have a whole number quotient. As with Student 1 and others, we should work with smaller quotients to develop understanding of the relationship between dividend, divisor, and quotient.

## TASK 4A: Which expression has the greatest quotient? Use models, numbers, or words to explain your thinking.

$$20 \div \frac{1}{4} \qquad\qquad 20 \div \frac{1}{2} \qquad\qquad 20 \div \frac{1}{5}$$

## Student Work 1

$$20 \div \frac{1}{2} \qquad 20 \div \frac{1}{4} \qquad 20 \div \frac{1}{5}$$

Which expression above has the greatest quotient?

Use models, numbers, or words to explain your thinking.

bigger

halves are bigger. There will be more of them because they take up more space

## Student Work 2

$$20 \div \frac{1}{2} \qquad 20 \div \frac{1}{4} \qquad \boxed{20 \div \frac{1}{5}}$$

Which expression above has the greatest quotient?

Use models, numbers, or words to explain your thinking.

There are 5 ⅕ in 1. There are 4 ¼ in 1 and 2 halves in 1 If there are more 5s in 1 there will be more in 20.

### Student 3

Student 3 shows reliance on procedure for finding mathematical solutions. Unfortunately, he applies the wrong procedure. In doing so, he arrives at an answer that is completely opposite of the correct answer. He also adds the statement "the bigger the number, the smaller the answer." This is either an observation he makes after completing the computation, or it is a misconception that he holds in general.

### Student 4

Student 4 reasons about the size of the pieces in one whole. He explains that there are more fifths in 1 than fourths or halves. His drawing justifies his statement. He extends this reasoning to more wholes by noting that there will be more fifths in 20 because there are more fifths in 1.

## USING EVIDENCE

*What would we want to ask these students? What might we do next?*

### Student 3

Sometimes students acquire procedures prematurely at home. In such cases, it is a good idea to communicate with home the challenges that Student 3 is having and the actions that we are taking. We may ask parents to avoid the procedure for a short time. It may also be helpful to use homework that practices other concepts for a time. Essentially, we need to separate Student 3 from these procedures for a time. It is clear that his reliance on them undermines his success. Like Students 1 and 2, we need to rebuild understanding of division with models and representations. We will connect these with equations. When doing so, it is especially important that our connections reinforce the meaning rather than the procedure associated with the equation.

### Student 4

Student 4 shows refined reasoning about division with fractions. Our next step is to work with reasoning with division of fractions other than unit fractions. We might provide a similar prompt that mingles unit fractions with other fractions to see if he naturally makes this connection. We will want to build this understanding with questioning and modeling if he doesn't. We will also make use of his representational understanding to make this connection.

$$20 \div \frac{1}{4} \qquad\qquad 20 \div \frac{1}{2} \qquad\qquad 20 \div \frac{1}{5}$$

## Student Work 3

$$20 \div \frac{1}{2} = \frac{20}{2} = 10 \qquad 20 \div \frac{1}{4} = \frac{20}{4} = 5 \qquad 20 \div \frac{1}{5} = \frac{20}{5} = 4$$

Which expression above has the <u>greatest quotient</u>?

$$20 \div \frac{1}{2} = \frac{20}{2} = 10$$

Use models, numbers, or words to explain your thinking.

① $\frac{20}{1} \div \frac{1}{2} = 10 \qquad 20 \div 2 = \boxed{10}$

② $\frac{20}{1} \div \frac{1}{4} = 5 \qquad 20 \div 4 = 5$

③ $\frac{20}{1} \div \frac{1}{5} = 4 \qquad 20 \div 5 = 4$

The bigger the number (□) the small the answer

10 (□) is bigger answer then 5 and 4

## Student Work 4

$$20 \div \frac{1}{2} = 10 \qquad 20 \div \frac{1}{4} = 5 \qquad 20 \div \frac{1}{5} = 4$$

Which expression above has the greatest quotient?

$$20 \div \frac{1}{2}$$

Use models, numbers, or words to explain your thinking.

10 is half of 20

10 is greater than 5 and 4

## OTHER TASKS

- What will count as evidence of understanding?
- What misconceptions might you find?
- What will you do or how will you respond?

 Visit this book's companion website at **resources.corwin.com/ minethegap/6-8** for complete, downloadable versions of all tasks.

**TASK 4B:** Danny knows $50 \times \frac{1}{2} = 25$. Because of this, she knows that $50 \times \frac{1}{4}$ must be less. Do you agree with Danny? What do you think the new product might be? Use pictures, numbers, or words to explain your thinking.

Understanding how the size of factors impact products is critical for determining the reasonableness of our answers. It is likely that some, if not most or even all, of our students believe that multiplication yields a product larger than the first factor, but as we know, this isn't the case. In this task, our students are asked to describe why a product of a factor and $\frac{1}{4}$ will be less than the product of the same factor and $\frac{1}{2}$. Our students are likely to reason that it makes sense because $\frac{1}{4}$ is less than $\frac{1}{2}$. It will be interesting to see if they recognize that the product will be 12.5 (or exactly half of 25). Some students may complete the computation to justify their solution. This may not be indicative of the reasoning we are seeking in our students.

**TASK 4C:** Oscar noticed these two columns of equations on the board.

$50 \div \frac{1}{10} = 500$  $\qquad$  $50 \div \frac{2}{10} = 250$

$40 \div \frac{1}{10} = 400$  $\qquad$  $40 \div \frac{2}{10} = 200$

$30 \div \frac{1}{10} = 300$  $\qquad$  $30 \div \frac{2}{10} = 150$

$20 \div \frac{1}{10} = 200$  $\qquad$  $20 \div \frac{2}{10} = 100$

 **MODIFYING THE TASK**

This task can be modified to examine patterns when computing other numbers. For example, we could explore the results of multiplying by $\frac{1}{10}$ and $\frac{2}{10}$ by simply changing the operation in the prompt.

**What patterns do you notice about the equations?**

**So how can knowing $80 \div \frac{1}{10} = 800$ help you solve $80 \div \frac{2}{10}$?**

*Intentional arrangement of equations can help our students see patterns in computations.* The quotients of the equations are provided so that students can focus on the relationships. We want our students to notice that the quotients of a number and $\frac{2}{10}$ are half of the quotient of the number and $\frac{1}{10}$. Though students may recognize the pattern, they may not be able to explain that they are dividing by twice as much so then the quotient will be half as much. The extension on the task is an opportunity to show that students can generalize their understanding.

**TASK 4D:** Which of the expressions below has a quotient greater than 10?

$$5 \div \frac{1}{3} \qquad\qquad 3 \div \frac{1}{3} \qquad\qquad 24 \div \frac{2}{3} \qquad\qquad 10 \div \frac{3}{4}$$

**Explain how you know.**

This task requires students to reason about a quotient and compare the result to a known. Students can convey their understanding with equations or drawings. Students may approach the expression by reasoning about dividing the dividend by the unit fraction. Then, they multiply that number by the number of unit fractions. For example, in $24 \div \frac{2}{3}$, students may note that $24 \div \frac{1}{3}$ is 72 so they double the 72 because there are *two* thirds. The latter shows a more refined sense of fractions. Through discussion, we can connect the procedures of some students with the representations of other students, and then with the fraction sense of others, to enable students to develop full understanding of the mathematics. We can discuss when some strategies are more efficient than others, though each will provide the same result.

NOTES

# BIG IDEA 5
## Problem Solving With Fractions

### TASK 5A

A bakery uses $2\frac{1}{2}$ cups of sugar for a cake.

The chart shows the number of cakes they make each day of the week.

How many more cups of sugar do they need on Friday than on Tuesday?

Use pictures, numbers, or words to explain your thinking.

| Days | Cakes Made |
|---|---|
| Sunday | 10 |
| Monday | 4 |
| Tuesday | 5 |
| Wednesday | 5 |
| Thursday | 6 |
| Friday | 9 |
| Saturday | 10 |

### About the Task

**MODIFYING THE TASK**

Using data tables is an efficient way to prompt a myriad of problems. This task can be easily modified by offering different questions. For example, four different prompts could be used for a station-based, collaborative, problem-solving lesson.

Making sense of fractions, especially computation with fractions, may be best done in context. *This collection of tasks provides multiplication of fractions in context.* This first task is not a traditional story problem. It replicates the types of problems we are more likely to encounter in the real world. In this problem, students gather data from a table, calculate with it, and compare the results.

### Anticipating Student Responses

Students will identify how they found the values for Friday and Tuesday. Some students will multiply each day by $2\frac{1}{2}$ before finding the difference. Other students may find the difference of cakes made between the two days and then multiply that amount by $2\frac{1}{2}$. Both approaches are reasonable, though the latter could be considered more efficient. Some of our students may

## PAUSE AND REFLECT

- How does this task compare to tasks I've used?
- What might my students do in this task?

Visit this book's companion website at **resources.corwin.com/minethegap/6-8** for complete, downloadable versions of all tasks.

make calculation errors. Students may convert $2\frac{1}{2}$ to an improper fraction before computing, while others will find partial products by multiplying by 2 and then by $\frac{1}{2}$. Other errors may occur because students don't understand what the problem is asking, because a key word is lacking, because there are multiple steps, or because they misread the table.

NOTES

### Student 1

**MINING TIP**

Avoid using key words as an instructional approach to problem solving. Doing so can create misconceptions about problems and set students up for incorrect solutions.

Student 1's work shows significant misunderstanding. Her work represents two different challenges students might have when working with this problem. *She subtracts $2\frac{1}{2}$ from 10 because she notes that the problem asks "how many more" and she connects this phrase with subtraction.* Her computation is also flawed. She says $\frac{10}{1} - 2\frac{1}{2} = 2\frac{9}{1}$. To make sense of this error, we can presume that she subtracts in either direction to find a result. In other words, she subtracted left to right for the numerator (10 − 1). She subtracts right to left for the denominator (2 − 1). It's likely that she simply brings the whole number over to her "difference."

### Student 2

Student 2 uses the more efficient strategy. She finds the difference of the two days (9 − 5). She then multiplies the difference by $2\frac{1}{2}$. She uses this strategy instead of multiplying both values by $2\frac{1}{2}$ and then finding the difference of the products. Her note about the "key" communicates that each number in the table is multiplied by $2\frac{1}{2}$. She doesn't multiply $4 \times 2\frac{1}{2}$ correctly. Instead, she multiplies 4 × 2 and then adds the half to the product yielding an inaccurate result of $8\frac{1}{2}$.

*What would we want to ask these students? What might we do next?*

### Student 1

This problem shows that Student 1 needs work with problem solving and computation. We can work on both concepts at the same time. First, we want to develop problem-solving strategies. We can have her work with models and drawings and connect these with equations. Computation work should revisit addition and subtraction of fractions. We would be wise to begin with fractions less than 1 before moving to mixed numbers. It would also be wise to work with less complicated denominators, such as halves, fourths, eighths, and twelfths.

**MINING HAZARD**

We must consider all of a student's work when determining his or her understanding. We can combine students' written thoughts with their diagrams or drawings to establish full understanding. However, additional ideas may not always link their ideas.

### Student 2

Student 2 shows proficiency with solving the problem. She has made sense of the problem and applies an efficient strategy. *Her writing doesn't fully explain why she subtracted 9 − 5.* Looking above, we may think she is linking the values to the associated days. But after looking closely, we see that she is making bonds of ten, which is not relevant to the task. Our next step with her is to revisit multiplication with fractions and mixed numbers. It may be a simple oversight that she did not include the half in her multiplication; on the other hand, it may be a more significant problem.

**TASK 5A:** A bakery uses $2\frac{1}{2}$ cups of sugar for a cake. The chart shows the number of cakes they make each day of the week. How many more cups of sugar do they need on Friday than on Tuesday? Use pictures, numbers, or words to explain your thinking.

## Student Work 1

A bakery uses $2\frac{1}{2}$ cups of sugar for a cake.

The chart shows the number of cakes they make each day of the week.

| Days | Cakes Made |
|------|------------|
| Sunday | 10 |
| Monday | 4 |
| Tuesday | 5 |
| Wednesday | 5 |
| Thursday | 6 |
| Friday | 9 |
| Saturday | 10 |

How many more cups of sugar do they need on Friday than on Tuesday?

$$\frac{10}{1} - 2\frac{1}{2} = 11$$

Use pictures, numbers, or words to explain your thinking.

I did $\frac{10}{1} - 2\frac{1}{2} = 2\frac{9}{1}$ I subtracted because the problem says how many more cups of sugar and $2\frac{9}{1}$ is an improper fraction $9+2=11$ and that is how I got my answer.

## Student Work 2

A bakery uses $2\frac{1}{2}$ cups of sugar for a cake.

The chart shows the number of cakes they make each day of the week.

| Days | Cakes Made |
|------|------------|
| Sunday | 10 |
| Monday | 4 |
| Tuesday | 5 |
| Wednesday | 5 |
| Thursday | 6 |
| Friday | 9 |
| Saturday | 10 |

How many more cups of sugar do they need on Friday than on Tuesday?

$$9-5=4 \times 2\frac{1}{2} = 8\frac{1}{2} \quad \text{Key } 2\frac{1}{2}$$

Use pictures, numbers, or words to explain your thinking.

I know they awnser because nine subtrack five eecal four. four times towand onehafe eecal eight and $\frac{1}{2}$ I think.

### Student 3

Student 3's strategy is logical. She identifies the number of cakes made on both days. She multiplies these values by $2\frac{1}{2}$. She then subtracts the products to find a difference. Her solution is incorrect. She multiplies each number (9 and 5) by $2\frac{1}{2}$ incorrectly. In each, she multiplies the whole numbers and ignores the half. She then adds the half back to the product.

### Student 4

Student 4 finds the solution. *She completes quite a few computations to find her solution.* It is fine that she multiplies both days by $2\frac{1}{2}$ before finding the difference. We can clearly see why she multiplied by 9 and 5 respectively as she connects the products with the days. We can also see how she found the products. She then shows her subtraction.

## USING EVIDENCE

*What would we want to ask these students? What might we do next?*

### Student 3

*We need to separate Student 3 from these procedures for a time.* It is clear that her reliance on them undermines her success. Like Students 1 and 2, we need to rebuild understanding of division with models and representations. We will connect these with equations. When doing so, it is especially important that our connections reinforce the meaning rather than the procedure associated with the equation.

### Student 4

Student 4 shows refined reasoning about division with fractions. Our next steps are to work with reasoning with division of fractions other than unit fractions. We might provide a similar prompt that mingles unit fractions with other fractions to see if she naturally makes this connection. We will want to build this understanding with questioning and modeling if she doesn't. We will also make use of her representational understanding to make this connection.

**TASK 5A:** A bakery uses $2\frac{1}{2}$ cups of sugar for a cake. The chart shows the number of cakes they make each day of the week. How many more cups of sugar do they need on Friday than on Tuesday? Use pictures, numbers, or words to explain your thinking.

## Student Work 3

A bakery uses $2\frac{1}{2}$ cups of sugar for a cake.

The chart shows the number of cakes they make each day of the week.

| Days | Cakes Made |
|------|------------|
| Sunday | 10 |
| Monday | 4 |
| Tuesday | 5 |
| Wednesday | 5 |
| Thursday | 6 |
| Friday | 9 |
| Saturday | 10 |

How many more cups of sugar do they need on Friday than on Tuesday?

$2\frac{1}{2} \times 9 = 18\frac{1}{2}$

$2\frac{1}{2} \times 5 = 10\frac{1}{2}$

$18\frac{1}{2}$
$-10\frac{1}{2}$
$8$ cups of sugar

Use pictures, numbers, or words to explain your thinking.

I got 8 wholes because it says How many more did Friday than tuesday. So I did $2\frac{1}{2}$ multiplied the days that where in the question and I got $18\frac{1}{2}$ and $10\frac{1}{2}$ so I subtracted those numbers and got my answer witch is 8 cups of sugar.

## Student Work 4

A bakery uses $2\frac{1}{2}$ cups of sugar for a cake.

The chart shows the number of cakes they make each day of the week.

| Days | Cakes Made |
|------|------------|
| Sunday | 10 |
| Monday | 4 |
| Tuesday | 5 |
| Wednesday | 5 |
| Thursday | 6 |
| Friday | 9 |
| Saturday | 10 |

How many more cups of sugar do they need on Friday than on Tuesday?

$2\frac{1}{2} = 1$ cake

$2\frac{1}{2} \times 5$

$2\frac{1}{2} \times 9$

$\frac{5 \times 1}{1 \times 2} = \frac{5}{2} = 2\frac{1}{2}$

$2 \times 5 = 10$

$\frac{1}{2} \times \frac{9}{1} = \frac{9}{2} = 4\frac{1}{2}$

$2\frac{1}{2} + 10 = 12\frac{1}{2}$ cups on tuesday

$2 \times 9 = 18$

$18 + 4\frac{1}{2} = 22\frac{1}{2}$ cups on friday

$22\frac{1}{2}$
$-12\frac{1}{2}$
$10$ cups of suger more

Use pictures, numbers, or words to explain your thinking.

## OTHER TASKS

- What will count as evidence of understanding?

- What misconceptions might you find?

- What will you do or how will you respond?

 Visit this book's companion website at **resources.corwin.com/ minethegap/6-8** for complete, downloadable versions of all tasks.

 **CONTENT CONNECTION**

As students work with application problems, they are building necessary foundational understanding that will be needed for later work with expressions and equations. By focusing on finding the unknown, look for formal and informal ways students are representing the situation.

**TASK 5B:** A construction company is paving 18 miles of road. It can pave $\frac{3}{4}$ of a mile in a day. How many days will it take to complete the road? The company says it will take 3 more days to complete a total of 20 miles. Do you agree or disagree with the company? Use models, numbers, or words to explain your answer.

*Task B is a problem with two different prompts.* In the first prompt, students find the quotient of $18 \div \frac{3}{4}$. Some students may simply note that $\frac{3}{4}$ is about 1, so the company can do about 1 mile per day. With this thinking, students will justify 18. Others may justify that it is a little more or a little less than 18 as they attempt to compensate for the fact that $\frac{3}{4}$ is not one whole. Some students may multiply thinking of an unknown factor, while others may multiply because they don't understand the context of the problem. In the second prompt, students have to reason about extending the road being paved.

 **MINING HAZARD**

Crafty, resourceful students who leverage understanding and patterns for "easy" completion of a task shouldn't be penalized. However, we do want them to be able to explain how they understand the patterns and relationships.

**TASK 5C:** Sets of mixed numbers have a difference of $\frac{3}{4}$. Identify different examples of these sets. Use models, words, or numbers to justify one of your solutions.

As we know, problem solving is not always relegated to word problems. Open-ended tasks such as this are great examples. It connects to ideas of decomposition of fractions or mixed numbers. Students will justify their solutions with models or calculations. *The intent of the task is for students to reason about the relationship between fractions while developing a sense of benchmarks.* Some students will take advantage of the "missing piece" in $\frac{3}{4}$ ($\frac{4}{4} - \frac{1}{4}$). Using this idea of a missing piece, they might then add something to both (e.g., adding 5 to each results in $6 - 5\frac{1}{4}$). Other students may make use of mixed number differences, creating an equation that reads $1\frac{3}{4} - 1 = \frac{3}{4}$ or $6\frac{3}{4} - 6 = \frac{3}{4}$.

**TASK 5D:** Complete a model for each expression. Write a word problem that could be solved with one of the expressions.

$$4 \div \frac{1}{4} = \qquad 5 \div 12 = \qquad 6 \div \frac{2}{3} =$$

*Word problems for expressions or equations support the development of problem-solving skills.* When students do this, they need to think about context, how problems are posed, how numbers relate to one another, and how questions in problems may be asked. We often do this with addition, subtraction, multiplication, or division of whole numbers. Applying this idea to multiplication or division with fractions can be considerably more challenging. Providing the models before the story supports the work in two ways. It helps us see if students can represent the division expression with a model, and it helps students make sense of the expression so that they can wrap a context around it.

**MODIFYING THE TASK**

The task features whole number quotients. It can be modified for any expression.

NOTES

# BIG IDEA 6
## Decimals as Numbers

### TASK 6A

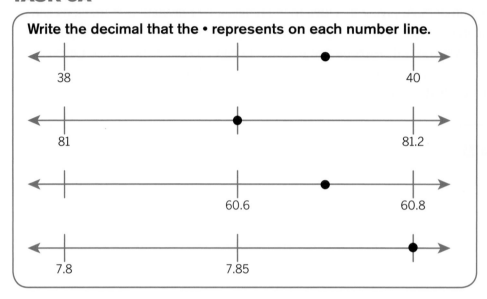

Write the decimal that the • represents on each number line.

38                    40

81                    81.2

60.6                 60.8

7.8          7.85

### About the Task

Conceptual understanding of decimals may occasionally overrely on grids and similar models. In doing this, we may unintentionally undermine the meaning of decimals as numbers. In other instances, we work with partitioned number lines that ask for recall of the value of locations, but deeper understanding may not be established. *In this task, our students have to identify a point on a number line by considering the relationship between endpoints that are whole numbers or decimals.* An unmarked endpoint is provided to help our students reason about the endpoints.

**MODIFYING THE TASK**

Change the values of the endpoints to align with the decimal place values featured in your mathematics class. Also consider labeling one endpoint and one midpoint instead of both endpoints.

### PAUSE AND REFLECT

• How does this task compare to tasks I've used?

• What might my students do in this task?

 Visit this book's companion website at **resources.corwin.com/minethegap/6-8** for complete, downloadable versions of all tasks.

## Anticipating Student Responses

Students with flawed understanding of decimals will offer a wide range of values. In all cases, the midpoint is useful for establishing the value of the point. Some students may also partition the space between midpoint and endpoint. In the first prompt, the midpoint represents 39, so the point represents a decimal halfway between 39 and 40. This reasoning may also be applied to the second number line with endpoints of 81 and 81.2. The last number line may be most challenging for students as it extends reasoning to hundredths. Look for a frequent misconception that 7.86 is the missing endpoint.

NOTES

### Student 1

Student 1 shows some interesting ideas about decimals. Three of his number lines are incorrect but for good reason. In the first number line, he estimates a value between 38 and the next whole number 39 instead of 40. In the second number line, he records the number halfway between 81 and 82. 81 is the starting point, and 81.2 may have been misread as 82. In the third number line, it seems that he begins with 60.6 and counts on adding five hundredths. This would be the mid-point between 60.6 (given) and the next tenth (60.7).

### Student 2

Student 2 has two correct and two incorrect number lines. We might note that he correctly identifies midpoints between two decimals in the second and third number lines. The first number line shows reasoning about a point that appears more than halfway between whole numbers. However, he doesn't consider the value of the second endpoint. The last number line shows inaccurate thinking about hundredths. He simply adds another hundredth instead of continuing the skip-count of five hundredths.

*What would we want to ask these students? What might we do next?*

### Student 1

Student 1's first response shows that he has some ideas about estimating points between two whole numbers. Even when errors, like this student's work, seem to clearly indicate a misread, we have to be careful not to assume why errors occur, especially in tasks like this that don't have students communicating their reasoning. We want to ask him to read the endpoints and describe the difference between endpoints. This will show if he simply misread the numbers or if the cause is more problematic. It seems likely that he misread the endpoints. Yet, his work with the third and fourth number lines indicates somewhat sophisticated thinking. Does it make sense that he would misread three different endpoints? We might also note that he recognizes that five hundredths is halfway between tenths. Our next step with Student 1 would be to revisit points on a number line with endpoints of regular endpoints (between two whole numbers). We can then move to greater differences between endpoints.

### Student 2

Student 2 shows that he can find midpoints with tenths. His first number line shows that he can relate a decimal placement relative to halfway (five tenths). This assumes that he misread or ignored the value of the second endpoint. His fourth number line shows that he doesn't have the same proficiency with hundredths. We can build from his accurate thinking. To do this, we can have him find points between two consecutive numbers. From there, we can extend the number line to the next consecutive number. We can have discussions about the numbers in the extended space. We will want to continue work with tenths before slowly moving to hundredths.

### Student Work 1

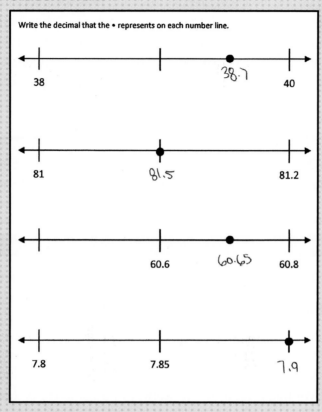

### Student Work 2

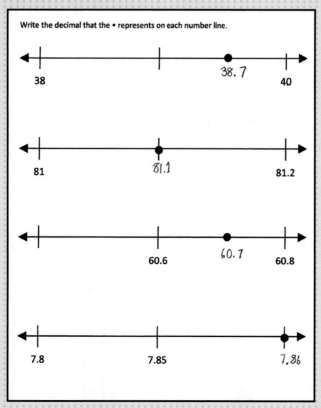

### Student 3

Student 3 offers interesting ideas about placing numbers on number lines. In the first number line, he clearly recognizes that 39 is the midpoint on the number line. He then knows that the point is a number between 39 and 40. His second number line is correct. His third number line is correct as well, but we might note that his left endpoint is incorrect. His last number line is inaccurate. It may be that it is just off similar to other work on the task. Or, it might be that he confuses 7.89 for 7.9.

### Student 4

Student 4 identifies three points correctly. Like other students in this sample, he identifies 38.5 instead of 39.5. He shows that he has ideas about midpoints between whole numbers. It is possible that he overlooks the second endpoint.

**USING EVIDENCE**

*What would we want to ask these students? What might we do next?*

### Student 3

Student 3 shows unique thinking that we will want to investigate further. In three of four number lines, his points or endpoints are slightly off. It may be attributed to the open number line. It may be that he doesn't reason about intervals and half-intervals. We might want him to work with a complete or tic-marked number line and look to see if he counts from point to point using single intervals. We may also work with folding strips of paper or sentence strips to model distances between values. From there, we want to skip intervals or change the space between noted intervals. Student 3 will also benefit from examining patterns in a string of decimals.

### Student 4

Assumptions about student work are mining hazards. However, we can look for patterns in responses to substantiate our assumptions. Student 4 is a good example. He shows in three of his number lines that he can find points accurately. It is likely that the first is an oversight of some sort. It is wise to give a few other prompts similar to the first to determine what the cause is. If there is a misconception, we will want to build from his understanding of points between consecutive values before extending the number line in either direction. As we do this, we want to highlight how the midpoint changes as we extend the number line.

## Student Work 3

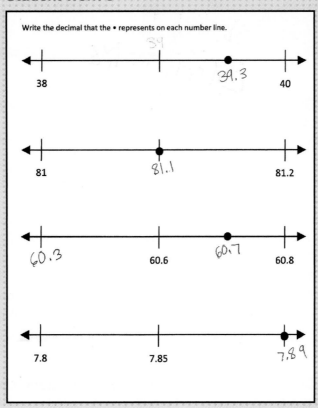

Write the decimal that the • represents on each number line.

## Student Work 4

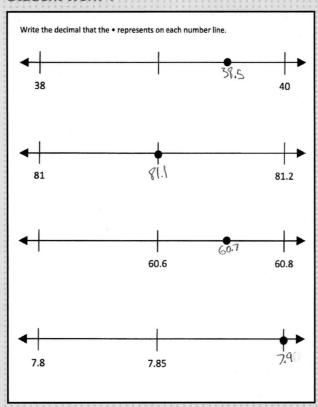

Write the decimal that the • represents on each number line.

- What will count as evidence of understanding?
- What misconceptions might you find?
- What will you do or how will you respond?

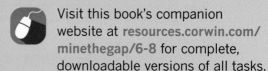

Visit this book's companion website at **resources.corwin.com/minethegap/6-8** for complete, downloadable versions of all tasks.

**TASK 6B:** Place 42.45 on the number line. Tell how you knew where to place 42.45 on the number line.

42

This task requires students to negotiate the relationship between tenths and hundredths knowing that hundredths are somewhere between two tenths. Proficient students may try at first to count by hundredths before reconsidering their strategy. Counting by tenths and then establishing 42.45 as the midpoint between 42.4 and 42.5 will be the most likely strategy. Some students may also reason that 42.45 is about 42.5 so the point is about "halfway" between 42 and 43.

**TASK 6C:** Write the missing numbers in the darker boxes ☐ on the decimal chart.

| | | | | | | | | | |
|---|---|---|---|---|---|---|---|---|---|
| 0.01 | 0.02 | 0.03 | 0.04 | | 0.06 | 0.07 | 0.08 | 0.09 | 0.10 |
| 0.11 | 0.12 | | | | | | | | |
| | 0.22 | | | | | | ☐ | | |
| | | | 0.34 | ☐ | | | | | |
| | | | ☐ | | | | 0.48 | | |
| 0.51 | | | | | | | ☆ | | |
| | | ☐ | ☐ | | | | | | |
| | | ☐ | ☐ | | | | 0.79 | | |
| | | | | | | | | | |
| | | 0.93 | | | | | | 0.99 | 1.00 |

**Tell how you know the value of the space with the star.**

This task makes use of a different take on a hundred chart that students work with in primary grades. That model showed the relationships between whole numbers in a sequence of 100. This chart features benchmarks and missing decimals. Students with developed understanding of decimals will use the benchmarks and count on or count back strategies making use of tenths. Some students may need to count by recording every hundredth in the empty spaces. Other students may not be able to complete the chart because they need additional work with decimals.

**TASK 6D:** **Break apart the number in the circle in three different ways.**

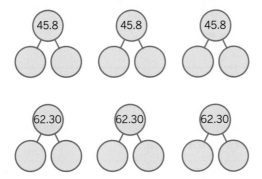

In this task, two different decimals are being decomposed. The first uses place value to the tenths while the second uses place value to the hundredths. Students will decompose in different ways. Again, we should look for students who rely on breaking the number (45.8) into a whole number (40) and a decimal (5.8) rather than decomposing into two different decimal numbers (41.4 and 4.4). The task can be used as a journal activity or station activity. After completing the task, we can bring the class together to share and chart examples. This exposes them to a wide range of decompositions and can improve their own thinking about decomposition.

NOTES

# BIG IDEA 7
## Addition and Subtraction With Decimals

### TASK 7A

Look at the numbers in the table.

| 3.4 | 6.05 | 17.21 | 12.91 |
|------|------|-------|-------|
| 14.76 | 5.85 | 10.46 | 8.50 |

Find a pair of numbers that has a difference of 2.45. Use models, numbers, or words to explain how you selected your numbers.

### About the Task

When subtracting decimals, like whole numbers, we may move too quickly to a procedure. Developing reasoning about subtraction develops our students' ability to determine if they carried out the procedure correctly. *In this task, students reason about the differences of decimals.* At first glance, it may seem like there are too many choices. However, if we reason about the size of the difference (2.45), we find that many of the pairs are not possible. For example, the difference between 3.4 and the numbers greater than 10 (14.76, 17.21, etc.) is clearly more than 2.45.

### Anticipating Student Responses

Some students may procedurally check for all of the possibilities for a specific number. Others may make a computational error. This error may result in a

**MODIFYING THE TASK**

Consider limiting the place value to tenths or extending it beyond hundredths depending on the decimal understanding of your students.

## PAUSE AND REFLECT

- How does this task compare to tasks I've used?
- What might my students do in this task?

Visit this book's companion website at **resources.corwin.com/minethegap/6-8** for complete, downloadable versions of all tasks.

misidentification of the minuend and subtrahend. Students will justify their solutions in different ways. Some students may add up while others subtract. We may find some students who rationalize the difference between some of the options and then procedurally prove their selections.

NOTES

### Student 1

Student 1 finds more than one combination with a difference of 2.45. She proves her solutions with written computations. Interestingly, she justifies three of the four combinations. She validates the combinations by connecting addition and subtraction of 2.45 in different equations. It is likely that she worked left to right, thus using addition first to find her solutions and then writing the related subtraction equation. She notes that her computation proves these combinations. We should note that her equations don't illustrate procedural computation as they don't show evidence of regrouping.

### Student 2

Student 2 finds a correct solution. She notes that the difference between 8.50 and 6.05 is 2.45. Her computation reinforces her solution. Unlike Student 1, she shows evidence of regrouping, which indicates that she didn't reason about the difference solely. Her explanation may cause concern. She speaks to looking at the "tens place" of two numbers that had a difference of two. It's possible that her mistake is an oversight. It's also possible that she misunderstands place value as related to the decimal point.

**USING EVIDENCE**

*What would we want to ask these students? What might we do next?*

### Student 1

Student 1 finds more than one combination with a difference of 2.45. She uses addition and subtraction. Her equations don't show evidence of regrouping. This may indicate that she computed on scratch paper, used a calculator, or selected numbers by reasoning about them. In other words, she was able to identify that $5.85 - 3.4 = 2.45$ because 5 is 2 more than 3. With this reasoning, she was able to generate equations unsure of their exact result. We should investigate how she found her solutions. We might even consider modifying the task so that one or two combinations can be considered in this way as the other numbers or distractors would be considerably farther apart. We can begin to shift her focus to reasoning or estimating differences before procedurally finding them.

### Student 2

We must first find out why Student 2 mentions the use of the "tens place" rather than the ones place. We may need to revisit the meaning of place values in decimal numbers. Assuming it was a simple lapse, our work should focus on estimating and discussing estimates prior to calculating. After discussing, we can find the exact answers and compare estimates with actuals. During discussion, it is important that students highlight whole number differences between numbers as well as decimal differences. Like Student 1, we may highlight the connection between addition and subtraction to support reasoning. We may also want to begin with decimals that make use of benchmarks or friendly numbers. This may include decimals of 0.25, 0.5, 0.75, and other tenths.

## TASK 7A: Look at the numbers in the table:

| 3.4 | 6.05 | 17.21 | 12.91 |
|-----|------|-------|-------|
| 14.76 | 5.85 | 10.46 | 8.50 |

Find a pair of numbers that has a difference of 2.45. Use models, numbers, or words to explain how you selected your numbers.

## Student Work 1

Look at the numbers in the box.

| 3.4 | 6.05 | 17.21 | 12.91 |
|-----|------|-------|-------|
| 14.76 | 5.85 | 10.46 | 8.50 |

Find a pair of numbers that has a difference of 2.45. Use models, numbers, or words to explain how you selected your numbers.

$$3.4 + 2.45 = 5.85 \quad | \quad 5.85 - 2.45 = 3.4$$
$$10.46 + 2.45 = 12.91 \quad | \quad 12.91 - 2.45 = 10.46$$
$$6.05 + 2.45 = 8.50 \quad | \quad 8.50 - 2.45 = 6.05$$

All of these have a difference of 2.45 when add and subtracted

## Student Work 2

Look at the numbers in the box.

| 3.4 | 6.05 | 17.21 | 12.91 |
|-----|------|-------|-------|
| 14.76 | 5.85 | 10.46 | 8.50 |

Find a pair of numbers that has a difference of 2.45. Use models, numbers, or words to explain how you selected your numbers.

$$\begin{array}{r} 8.50 \\ - 6.05 \\ \hline 2.45 \end{array}$$

When I looked at the numbers I looked for two numbers where the tens place has a difference of $2 then I saw 6 and 8 so I tried those two numbers. When I was done subtracting I got a diffrence of 2.45

### Student 3

Student 3 finds that 8.50 and 6.05 have a difference of 2.45. Like others, she calculates to verify her solution. Her writing offers a different perspective. She talks about the relationships she knows. Specifically, she says that she knows the difference between 8 and 6 is 2 and that the difference between 50 and 05 is 45. She then notes that she combined these ideas to make her selection.

### Student 4

Student 4 adds 2.45 to a number to justify her solution. She notes that she begins with one of the options (10) and adds 2 to it. She explains how she knows that the sum of 46 + 45 is 91. We can see that her computation decomposes these addends by their place value. She then provides the equation 10.46 + 2.45 = 12.91.

**USING EVIDENCE**

*What would we want to ask these students? What might we do next?*

### Students 3 and 4

Student 3's calculation may lead us to overlook her consideration of partial sums or differences. It's important to note that she looks for whole number (8 − 6) and decimal (50 − 05) relationships to find her solution. Student 4's work is included to show evidence of students connecting addition and subtraction. She speaks to how she added up from one whole number before adding up from one decimal to another. Students 3 and 4 both show strategies for finding solutions to these sorts of tasks. We should continue to develop mental mathematics skills through reasoning and estimation. It can be challenging to generate lots of examples for use in this type of task. We might consider having students generate decimals for comparison by providing this task with a blank table and a new target difference. We can then use their created prompts as part of centers or opening routines.

**TASK 7A:** Look at the numbers in the table:

| 3.4 | 6.05 | 17.21 | 12.91 |
|-----|------|-------|-------|
| 14.76 | 5.85 | 10.46 | 8.50 |

Find a pair of numbers that has a difference of 2.45. Use models, numbers, or words to explain how you selected your numbers.

## Student Work 3

Look at the numbers in the box.

| 3.4 | 6.05 | 17.21 | 12.91 |
|-----|------|-------|-------|
| 14.76 | 5.85 | 10.46 | 8.50 |

Find a pair of numbers that has a difference of 2.45. Use models, numbers, or words to explain how you selected your numbers.

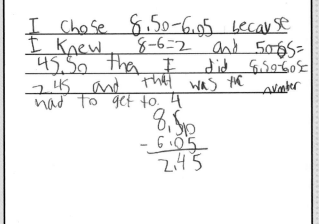

I chose 8.50-6.05 because I knew 8-6=2 and 50-05= 45.So then I did 8.50-6.05 = 2.45 and that was the number had to get to.

$$\begin{array}{r} 8.50 \\ -6.05 \\ \hline 2.45 \end{array}$$

## Student Work 4

Look at the numbers in the box.

| 3.4 | 6.05 | 17.21 | 12.91 |
|-----|------|-------|-------|
| 14.76 | 5.85 | 10.46 | 8.50 |

Find a pair of numbers that has a difference of 2.45. Use models, numbers, or words to explain how you selected your numbers.

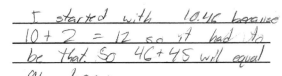

10.46 12.91

10.46 + 2.45 = 12.91

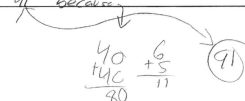

I started with 10.46 because 10 + 2 = 12 so it had to be that. So 46 + 45 will equal 91 because

$$\begin{array}{cc} 40 & 6 \\ +40 & +5 \\ \hline 80 & 11 \end{array}$$

$$\begin{array}{r} 80 \\ +11 \\ \hline 91 \end{array}$$

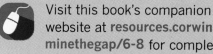 Visit this book's companion website at **resources.corwin.com/ minethegap/6-8** for complete, downloadable versions of all tasks.

**TASK 7B:** Subtract the decimals. Use a number line to show your work.

**10.5 − 6.3**   **22.75 − 20.50**   **30.5 − 14.5**

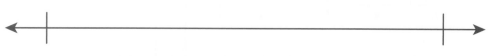

**MINING TIP**

**Some students only think of subtraction as take away. As we know, it can also mean the difference between two values. Number lines can develop and reinforce the latter.**

*This task asks students to make use of a number line to model subtraction with decimals. It allows students to make use of addition (adding up) or subtraction to find the difference between the two numbers.* Students may apply their understanding of decomposition to make one or both numbers easier to work with. For example, students might decompose 6.3 when subtracting it from 10.5. In this example, they may jump back 6 from 10.5 to land on 4.5. Then, they can make a jump of 0.3 backwards. Other students may prefer to take away decimal parts first followed by whole numbers, while other students will add up from the subtrahend to the minuend. When adding up, students are likely to rely on jumps of friendly numbers.

**TASK 7C:** Jackson says that when you add decimal numbers you have to add the decimals first. Do you agree with Jackson? Use models, numbers, or words to explain why you agree or disagree.

Student responses to this task are likely tied to the strategies they prefer to use when adding decimals. Students who rely on a standard algorithm are likely to agree that you have to add decimals first. In these cases, students may often add correctly but have some misconception or underdeveloped understanding of the meaning of addition with decimals. Some who prefer the algorithm may also recognize that you can decompose and add decimals in various ways. Other students who further developed computational strategies will also note decomposition or that you can start with any place value so long as you add like place values.

**TASK 7D:** There is a decimal missing from the first two numbers in the problem 47040 − 2931 = 17.73. Place the missing decimals. Use models, numbers, or words to explain how you knew where to place the decimals.

This task is a take on missing decimal or missing digit activities. It requires students to consider the size of the difference and determine the value of the subtrahend and minuend. Some students may place a decimal point to create hundredths in both in the subtrahend and minuend. Their reasoning may be that the difference is to the hundredths place so the other two values must be to the hundredths place as well.

In fact, that is why the zero in the thousandths place is provided in the minuend. Accurate students may try different combinations until they find the solution. These students are likely to use an algorithm or similar approach to justify their thinking. Other students will reason about the resulting difference as the decimal place is moved. For example, if the problem becomes 470.40 – 29.31, the difference will be greater than 400. If the problem becomes 47.040 – 2.931, the difference will be greater than 40.

NOTES

# BIG IDEA 8
## Multiplication and Division With Decimals

### TASK 8A

> Tell why you agree or disagree with each statement.
>
> $16 \div 8.5$ is less than 2
>
> $14 \div 3.5$ is less than 5
>
> $25 \div 4.75$ is about 5
>
> $47 \div 12.5$ is about 4

**MODIFYING THE TASK**

Modify the division expressions and conditions to align with the computation you are reinforcing. We can have students create prompts for use with the class as they become more comfortable with these types of tasks.

**MINING TIP**

Prompts for reasoning do not always provide consistent results. They can be difficult for developing mathematicians. The second prompt in this task may be more challenging than the last two.

### About the Task

*In this task, students reason about the results of equations.* For example, they consider if the quotient of $16 \div 8.5$ will be more or less than 2. *Even for adults, this type of reasoning can be quite challenging.* As noted throughout this book, reasoning is essential for accurate calculations. Working with this and similar tasks on a consistent basis helps students develop number sense and fluency. It also supports procedural understanding. Another idea for getting the most from these types of tasks is to have students make a prediction relative to the condition and then find the actual result. After doing so, we can compare estimations and actuals.

### Anticipating Student Responses

In most cases, students will use the whole number to make sense of the results. In division settings such as this, students are likely to think multiplication to

### PAUSE AND REFLECT

- How does this task compare to tasks I've used?
- What might my students do in this task?

Visit this book's companion website at **resources.corwin.com/minethegap/6-8** for complete, downloadable versions of all tasks.

find their solution. For example, the first prompt asks if $16 \div 8.5$ will be more or less than 2. Students might reason that $2 \times 8$ is 16, so we would multiply 8.5 by a number less than 2 to arrive at 16. Some students will compute using algorithms to establish their results.

NOTES

### Student 1

Student 1 has four accurate answers. All of her answers are derived from long division with decimals. She then compares her quotient to the prompt, but this isn't complicated once the exact quotient is found.

### Student 2

Student 2 explains that she doesn't have to compute the first prompt to confirm her thinking. Yet, her explanation talks more about the idea that $16 \div 8.5 \neq 2$ rather than the idea that $16 \div 8.5$ is less than 2. She writes a division equation for other problems before using multiplication to find an exact answer that she can compare to the situation presented in the prompt.

## USING EVIDENCE

*What would we want to ask these students? What might we do next?*

### Student 1

Student 1 shows preference for procedural approaches to comparison tasks. This isn't a deficiency. But in many cases, it can be inefficient. She will benefit from opportunities to estimate quotients or products of whole numbers without computation before applying these ideas to computation with decimals. We may also take advantage of number discussions and other mental exercises described for the other students in this task.

### Student 2

We can build on Student 2's understanding of the relationship between multiplication and division. Her reasoning is sound. However, she does rely on computations to justify her thinking. We want to help her make use of her understanding. We also want to help her shift from exact computations to estimation. For example, in the third prompt she multiplies $4.75 \times 5$ exactly. But it would seem that proof can be offered by knowing that $5 \times 5$ is 25 and that 4.75 is about 5.

## TASK 8A: Tell why you agree or disagree with each statement.

16 ÷ 8.5 is less than 2

14 ÷ 3.5 is less than 5

25 ÷ 4.75 is about 5

47 ÷ 12.5 is about 4

## Student Work 1

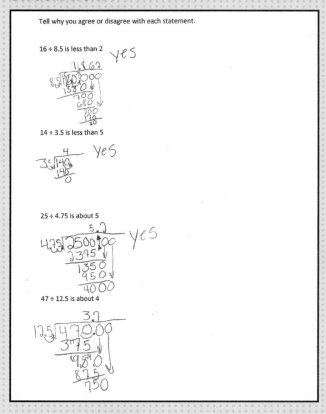

Tell why you agree or disagree with each statement.

16 ÷ 8.5 is less than 2    Yes

14 ÷ 3.5 is less than 5    Yes

25 ÷ 4.75 is about 5    Yes

47 ÷ 12.5 is about 4

## Student Work 2

Tell why you agree or disagree with each statement.

16 ÷ 8.5 is less than 2

I agree because
8.5 times 2 equals
17 not 16.

14 ÷ 3.5 is less than 5

I disagree
because 3.5 times
5 is 17.5 which is
greater than 14 so it
has to be less than.

25 ÷ 4.75 is about 5

I agree
because 4.75 times
5 is 23.75 which
is close to 5.

47 ÷ 12.5 is about 4

I agree
because 12.5 times
4 is 5 which is
about 4.

### Student 3

Student 3 is accurate in all four prompts. She explains her reasoning in the first prompt without actual computations to justify her accuracy. In the last three prompts, she provides the complete computation. In each of those three prompts, she uses multiplication to find a product. She then compares the product to the intended outcome. It is impressive that she accurately navigates the second prompt using multiplication. In that exercise, she makes sense of the size of factors. She relates that a greater product (17.5) is generated by greater factors ($3.5 \times 5$), so then the lesser product in the prompt (14) is caused by a factor less than those in the prompt ($3.5 \times 5$).

### Student 4

Student 4 connects division and multiplication to reason about the results of each prompt. She notes that $16 \div 8 = 2$, so 16 divided by something more than 8 will produce a lesser quotient. She also reasons with division in the second and third prompts. Her second response shows misunderstanding of how we can think about the divisor. Also note that she makes use of multiplication in the fourth response. We can infer that her reasoning for this prompt is based on the idea that $12.5 \times 4 = 50$ *and 50 is close to 47.*

*What would we want to ask these students? What might we do next?*

### Student 3

Student 3 has some good ideas about reasoning. Her paper/pencil computation supports her accuracy. We want to develop mental computation skills with Student 3 and others like her. We can do this by having number discussions about similar prompts. During these conversations, we should remove paper/pencil and other tools so that she can apply mental math and computations. We can then record her thinking or allow her to do so after communicating her strategy or approach. Doing so helps confirm accuracy and develop confidence with mental computation.

### Student 4

Student 4 demonstrates sound reasoning for three of the four prompts. She shows that she can make use of both multiplication and division. She also shows that her reasoning isn't always mathematically accurate, as evidenced in the second prompt. Like many other students, she needs more opportunities to apply her thinking to similar prompts. We should vary the prompts and be sure to debrief the strategies used. In certain situations, like the second prompt, we need to pose questions that expose the inaccuracies and promote alternative approaches.

16 ÷ 8.5 is less than 2

14 ÷ 3.5 is less than 5

25 ÷ 4.75 is about 5

47 ÷ 12.5 is about 4

## Student Work 3

Tell why you agree or disagree with each statement.

16 ÷ 8.5 is less than 2

I disagree. I don't even have to solve 8 can go into 16 twice but there is a half left over.

14 ÷ 3.5 is less than 5

I agree, when I solved the answer was 4 which less than five

25 ÷ 4.75 is about 5

I disagree, 6 is greater than 5, it may be close but to me about 5 is a hundreth place away.

47 ÷ 12.5 is about 4

I disagree, 50 is 3 numbers away from 47.

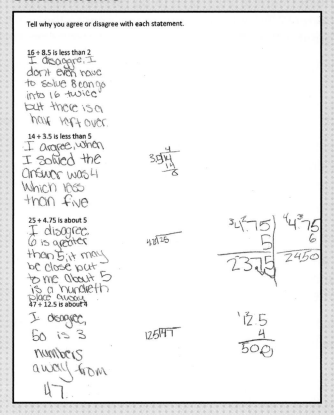

## Student Work 4

Tell why you agree or disagree with each statement.

16 ÷ 8.5 is less than 2

I agree because 16÷8=2 so somthing over 8 will be less than 2

14 ÷ 3.5 is less than 5

I disagree because 14÷2=7 so if you add 1.5 to 2 it would = 5.5

25 ÷ 4.75 is about 5

I agree because 25÷5=5 and 4.75 ist close to 5

47 ÷ 12.5 is about 4

I agree becaus 12.5 x4 is 50

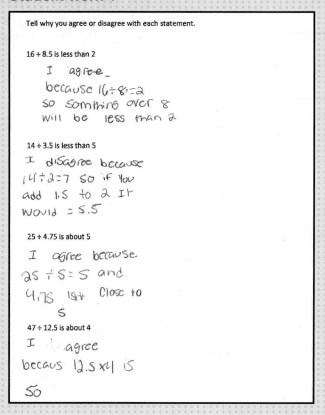

## OTHER TASKS

- What will count as evidence of understanding?
- What misconceptions might you find?
- What will you do or how will you respond?

Visit this book's companion website at **resources.corwin.com/minethegap/6-8** for complete, downloadable versions of all tasks.

**TASK 8B:** Annie found the quotient of 20 ÷ 0.75 to be greater than 20. Is it possible that Annie is correct? Tell why you agree or disagree that Annie's quotient would be greater than 20.

This task is a prompt that conflicts with students' misconceptions that division yields a quotient less than the dividend. Students may multiply 0.75 by 10 (twice) or by 20 to show that the product is less than 20. With this approach, students will rationalize that the quotient must be greater than 20 because 0.75 × 20 creates a product less than the dividend. Other students may note that 20 ÷ 1 is 20 and that 0.75 is less than 20. They will argue that a divisor less than 1 creates a quotient larger than the dividend. Some students may justify their thinking with smaller numbers and then extend their reasoning to the larger dividend of 20.

### CONTENT CONNECTION

Area models are critical to developing student conceptual understanding of multiplication. As students progress with expressions and equations, these models will help them understand more complex multiplication studied in algebra, including the extended distributive property.

**TASK 8C:** Emmett used an area model to multiply 3.6 × 14. Use an area model to show how Emmett might have broken apart the factors and multiplied. What product did Emmett find? Explain how you know.

*This task connects multiplication of decimals with the area model we have used for multiplying multi-digit whole number factors.* This model helps us develop partial product strategies as we can decompose numbers by place value. In this task, we might decompose 3.6 into 3 and 0.6 while decomposing 14 into 10 and 4. We don't always have to decompose numbers by their place value. Sometimes it's more convenient to break an even number in half (14 into 7 and 7). Doing so in this problem would yield 7 × 3 and 7 × 0.6. We would then double these partial products for the final product.

### MODIFYING THE TASK

Change the factors so that rounding produces a result contrary to the actual computation. This will lead to a discussion about estimation and reasoning versus rounding. For example, we could ask if 35 > 7 × 5.25. Students who round will say that it isn't because 7 × 5 equals 35!

**TASK 8D:** Tell if each is true or false. Explain how you found your answer without finding the exact value.

$$35.00 > 7 \times 5.50 \qquad 24.50 < 6.75 \times 4$$

Students with incorrect answers are likely to ignore the decimal portion of the factor. Others may think that multiplying by a decimal reduces the size of the product, as does multiplying a whole number by a decimal. For example, 7 × 0.5 produces a product less than 7. *Students with accurate thinking may round the decimal factor.* That strategy will work in this task but may not always. Some students will decompose the factor to create partial products. Other students will reason about the whole number factors and how the additional decimal product impacts the final product. Still others will actually compute even though the directions state the opposite.

**CHAPTER 3**

# INTEGERS

**THIS CHAPTER HIGHLIGHTS HIGH-QUALITY TASKS FOR THE FOLLOWING:**

- Big Idea 9: Representing Integers

  Integers are numbers that can be represented with different types of models. These models help us better understand the meaning of integers. Decomposing integers makes comparison and computation more efficient.

- Big Ideas 10 and 11: Representing Integers on Number Lines

  Integers can be represented on number lines to see relationships between quantities. We can also use number lines to compute with integers.

- Big Idea 12: Comparing Integers

  Different models can be used to compare integers. We can also compare integers by putting them into context and comparing them relative to zero.

- Big Idea 13: Addition With Integers

  Addition of integers is more than a mechanical process or set of rules. It is grounded in understanding and representation. Reasoning about addition with integers helps us make sense of our sums.

- Big Idea 14: Subtraction With Integers

  Subtraction of integers is more than a mechanical process or set of rules. It is grounded in understanding and representation. Reasoning about subtraction with integers helps us make sense of our differences.

- Big Idea 15: Multiplication With Integers

  Multiplication of integers is more than memorizing rules like a positive times a negative is a negative. We can represent multiplication with integers and examine the patterns of integer products to understand these rules.

- Big Idea 16: Division With Integers

  Division with integers is more than memorizing rules like a negative divided by a negative is a positive. We can represent multiplication with integers and examine the patterns of integer products to understand these rules.

# BIG IDEA 9
## Representing Integers

### TASK 9A

Break apart −18 in six different ways using the number bonds below.

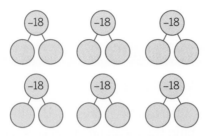

## MODIFYING THE TASK

The integer in the number bond can be modified to align with the values with which students are working. It might include integers between −10 and 10 or −100 and 100.

## About the Task

We can represent integers in many different ways. Number lines and integer chips are familiar choices. We often use contexts of elevation, money, temperature, and even yardage from football games. It's also important for our students to consider and represent integers in various decompositions. Doing this supports developing number sense with integers. *In this task, students are asked to decompose −18 in six different ways.* It builds on their experience decomposing whole numbers, decimals, and even fractions.

## Anticipating Student Responses

Our students may be comfortable representing integers with models or contexts. This task challenges them to think of them as the result of a number bond. Students with misconceptions will obviously create bonds that don't combine to

## PAUSE AND REFLECT

- How does this task compare to tasks I've used?
- What might my students do in this task?

Visit this book's companion website at **resources.corwin.com/minethegap/6-8** for complete, downloadable versions of all tasks.

make –18. Some students will show a combination of two negative integers such as –17 and –1, –16 and –2, or –10 and –8. Other students will show this combination as well as –18 represented by combining a positive and negative integer. Their examples might include –20 and 2 or –28 and 10.

NOTES

## Student 1

Student 1 seems to confuse the idea of number bonds with factor trees. He doesn't decompose −18 into two integers. Instead, he offers factor pairs. To confirm, we also see a graphic for factor pairs in the upper right-hand corner. At first glance, we might dismiss the work as a simple misunderstanding. But if we look closer, his combination of integers is flawed in the top row.

## Student 2

Unlike Student 1, Student 2 shows that he understands the idea of number bonds and decomposing integers. We should notice that each decomposition features two negative numbers. We should also notice that there is a distinct relationship between bonds. For example, −12 and −6 are a bond, and −2 and −16 are another. These two decompositions have a clear connection. He replicates this with −15 and −3, and −13 and −5. We might also note that every decomposition features at least one single-digit partial.

*What would we want to ask these students? What might we do next?*

## Student 1

This task is intended to determine if students can decompose an integer into two or more integers. We should first help Student 1 make sense of decomposition. The number bonds may be unfamiliar, or he may be muddling different ideas. It may be that he is able to decompose and this redirection will uncover it. If he isn't able to decompose −18, we will want to work with him to decompose integers. It may make the most sense to begin with positive integers followed by negative integers. We may first decompose the integer into two like-signed parts (negative decomposed into two negatives). In time, we would move to decomposing integers into a combination of positives and negatives. Along the way, we should be sure to use diverse models to represent the concept.

## Student 2

Student 2 shows understanding that we can build on. He shows that he understands and is able to decompose a negative integer into two negative integers. We should work to extend his understanding to other combinations, including combinations of negative and positive integers. Like Student 1, Student 2 will be better able to develop these compositions with support from various models, including integer chips and number lines. It would also be wise to record the combinations in ways to help students such as this to see the patterns within the combinations.

Break apart −18 in six different ways using the number bonds.

## Student Work 1

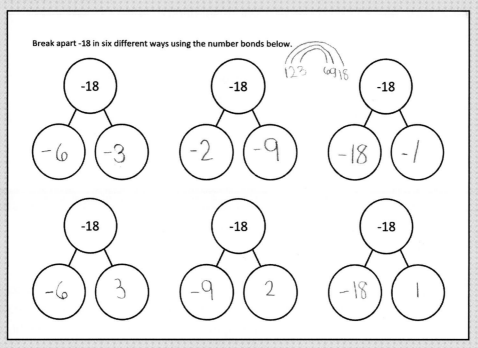

## Student Work 2

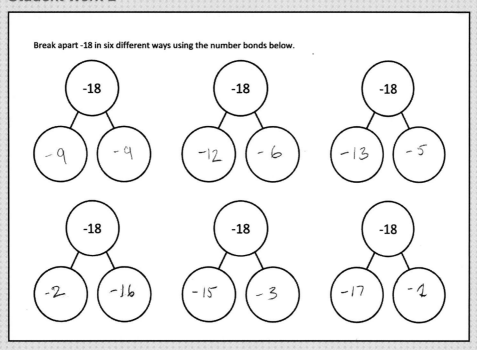

### Student 3

Student 3 decomposes −18 in clearly different ways. We see four combinations of negative integers. We also see a student who realizes that an integer can be decomposed into a bond of negative and positive numbers. We see this with −20 and 2. However, −13 and 5 show a clear misunderstanding.

### Student 4

Student 4 decomposes −18 in six distinct ways. None of the decompositions feature similar numbers. Four of the six decompositions feature pairs of negatives. The other two decompositions show correct combinations of a negative and a positive. One of them (−30 and 12) features two two-digit integers.

*What would we want to ask these students? What might we do next?*

### Student 3

Student 3's work is encouraging because he shows that he can decompose with two like-signed integers. He also shows that a negative and a positive can be combined. His error is cause for deeper investigation. It may be a simple oversight, or it may indicate that he is still developing an understanding of decomposition of integers and/or addition and subtraction of integers. Our next step with Student 3 is to revisit decomposition of integers with two different signs. We may start with single-digit integers and decompose them in different ways to discuss the similarities between the decompositions. We can then move to multi-digit integers.

### Student 4

Student 4 shows understanding of decomposition. He is able to do so with two like-signed integers as well as a bond of a positive and a negative. We can extend his work in different ways. We can begin to work with numbers greater than 50 or less than −50. We might consider three-digit integers or numbers with decimals. We may also begin to decompose integers into more than two parts. For example, −18 could be decomposed into −20, 1, and 1 or −40, 20, and 2.

## Student Work 3

Break apart -18 in six different ways using the number bonds below.

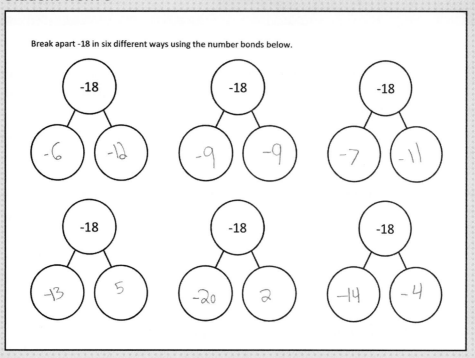

## Student Work 4

Break apart -18 in six different ways using the number bonds below.

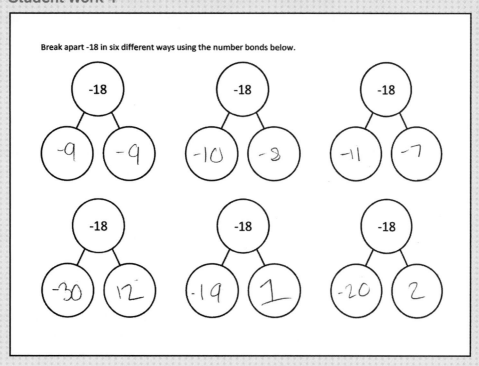

# OTHER TASKS

- What will count as evidence of understanding?

- What misconceptions might you find?

- What will you do or how will you respond?

 Visit this book's companion website at **resources.corwin.com/ minethegap/6-8** for complete, downloadable versions of all tasks.

 **MINING TIP**

Recording patterns in intentional ways helps students see the relationships between the values. Our tasks can feature these arrangements. It is also important that we do the same during class discussions.

 **MODIFYING THE TASK**

We might modify the task to have students place −12 on two different number lines. This would help us determine how well −12 relates to other integers.

A complete integer chart is available online at **resources.corwin. com/minethegap/6-8**.

 **CONTENT CONNECTION**

Real-world contexts help to build fluency with integer operations. This fluency will be critical as students expand understanding to representing and solving real-world problems using expressions and equations.

**TASK 9B:** Represent −12 in four different ways (on a number line, with two different real-world examples, and as an absolute value) on the Frayer model.

*Representing integers in different ways develops our students' understanding.* It is critical that they have opportunities to connect the different representations. This task is a Frayer model asking students to represent an integer in different ways. *We should look to see if −12 is reasonably placed between two endpoints.* The endpoints might feature −20 and 0, −20 and 20, or possibly −13 and −11. It's also possible that −12 could be an endpoint. We should also look for two different real-world contexts. Students who share two different elevations (e.g., sea level and elevators) are correct. However, the similar context may show a limited understanding of what integers can describe.

**TASK 9C:** Write the missing numbers on the integer chart.

Our students have experience working with number charts, including hundred charts and decimal charts. These charts help them represent and understand relationships between numbers. *This task features an integer chart.* Some integers are provided. To complete it successfully, some students may need to write every missing integer. More advanced students will count on from given or known numbers. Students may count on or count back by ones. Others may count on or back by tens moving vertically on the chart. For example, 10 less than 5 is −5 (or one box up), whereas 10 more than 50 is 60 (or one box down).

**TASK 9D:** Identify three different real-world examples of integers. Use pictures, numbers, or words to describe how each is an example of integers.

*As we know, there are many contexts for representing integers.* This includes elevation, temperature, bank accounts, yardage in football, distance, weights, and so on. This task asks students to identify different real-world examples of integers. It is a good task to provide to partners or small groups before bringing the class together to share ideas. We should look to see a range of examples. An abundance of or missing examples may be indicative of models that we have favored or neglected during instruction or within problems. We should also look to see if students recognize the relationship within the example relative to an origin or zero. We might also look to see if any of the examples are represented with number lines.

# BIG IDEA 10
## Representing Integers on Number Lines

### TASK 10A

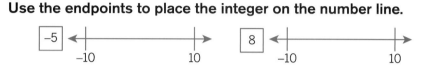

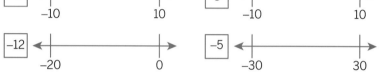

Use the endpoints to place the integer on the number line.

−5    −10    10      8    −10    10

−12    −20    0      −5    −30    30

**MODIFYING THE TASK**

Any integer and any endpoints can be used with this task. It may be interesting to see the same integer in all four boxes with four number lines, each featuring different endpoints.

### About the Task

*Representing integers on number lines establishes the relationships between integers and supports the development of our students' number sense.* It also lays the groundwork for computation with integers and graphing on coordinate planes. In this task, students are asked to consider how the integer in each box relates to the integer endpoints on the corresponding number line.

### Anticipating Student Responses

Students may create a tick mark for every integer between the endpoints. They may then use those to count on or count back from an endpoint to locate the given integer. These students may even label each tick mark, though that's not necessary. Some students may first locate zero and disregard counting on a specific side of zero (e.g., the positive side for the first number line). Other students may reason about the relationship between the given integer and the endpoints. They may locate zero and then reason that −5 is halfway between 0 and −10. A similar strategy may be applied to others, though they will not fall exactly halfway between any benchmark on the number line.

### PAUSE AND REFLECT

- How does this task compare to tasks I've used?
- What might my students do in this task?

 Visit this book's companion website at
resources.corwin.com/minethegap/6-8
for complete, downloadable versions of all tasks.

### Student 1

Student 1's work is evidence of misunderstanding. *Her challenges may extend beyond integers.* Her first response shows −5 as halfway between −10 and 10. This may be connected to her thoughts about 5, 50, 500, or even 12 as being halfway between two numbers. Her second number line may show that she has counted on from zero by reversing the endpoints. Her third number line is accurate. Her fourth may be evidence of counting on from an endpoint that she may perceive to be zero.

### Student 2

Student 2's tick marks are evidence of counting to find relationships. It also shows that she is relying on one-to-one counting rather than group or skip counting. But there is more. Her second number line shows counting by ones, but if we look more closely, it also seems to show that −10 is fairly past the midpoint of −20 and 0. It seems that −12 (or possibly −11) is halfway between −20 and 0. This may show that she doesn't recognize or consider −10 to be half of −20.

*What would we want to ask these students? What might we do next?*

### Student 1

We can see that Student 1 has unfinished learning. But there are different next steps that we might take. Her first line shows misconceptions about midpoints. For that, we should be sure to use number lines with diverse end- and midpoints. The second may indicate an oversight, though it may be more closely linked with the fourth number line in which she appears to be counting. In either case, we should work to help her develop understanding of benchmark relationships between numbers. We should work with halves and quarters of numbers (when possible) and show them on a number line. We should also consider more general relationships such as how 4 relates to 0 and 100, 0 and 10, 0 and 5, or −5 and 5.

### Student 2

One-to-one counting is a strategy for finding placements on number lines. As we know, it can be inefficient and inaccurate. Our greater goal is to develop reasoning about number or integer relationships. Student 2 shows that she finds the position of an integer but may not understand the relationship. As with Student 1, we should work with Student 2 to help her determine these relationships. It is possible that the understanding is developing with integers but already in place with other numbers. The fact that she discounts numbers to the right of zero in the first and third number lines may be an indication that this understanding is forming for integers.

## Student Work 1

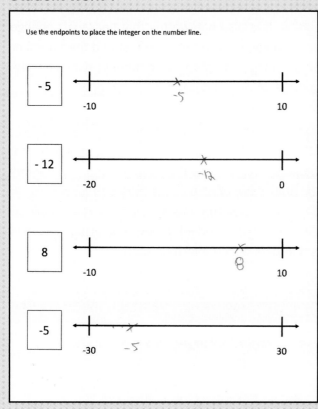

## Student Work 2

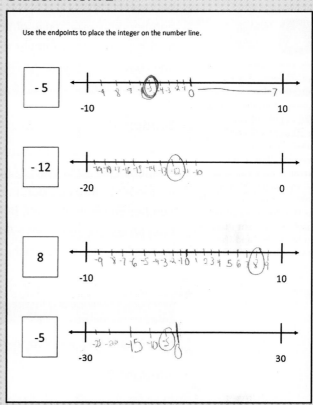

### Student 3

Student 3's work is similar to that of Student 2. She clearly uses tick marks to indicate one-to-one counting. Unlike Student 2, her midpoints are reasonable. She also shows that she needs to identify specific tick marks along the way. This may be for accuracy or it may be indicative of confidence or a need to actually see other numbers to make sense of positioning. It is also noteworthy that her labels change as the endpoints or integer to place change. For example, the first and third number lines have the same endpoints, but the integers to place are different (−5 and 8). The fourth number line features endpoints that are considerably farther apart.

### Student 4

Though not completely accurate, Student 4 shows that she is thinking of benchmark relationships when placing her integers. She reasons that 0 and −10 are midpoints for the first three number lines. *The placement of zero as the midpoint for the fourth number line is slightly off, but we might assume that it is an accuracy issue more so than an issue of understanding.* Like Student 3, Student 4 may need visual supports to make sense of relationships as numbers become farther apart.

**MINING HAZARD**

Inference and assumption can be pitfalls when making sense of student work. We might witness a student's correct thinking in class and assume that a new error is carelessness or oversight. We might infer or give the benefit of the doubt when in fact the student doesn't fully understand the concept.

*What would we want to ask these students? What might we do next?*

### Student 3

Student 3's work is a wonderful snapshot of how a student thinks about number relationships and how their understanding is developing. The first and third number lines feature the same endpoints, but her strategy for finding the placements of the integers is different. She shows that she sees a relationship between −5, −10, and 10. Yet, 8 isn't as comfortable, and so she relies on exact counting. She also shows that she can make use of benchmark relationships on the second and third number lines. We should continue to give her opportunities to work with and discuss how integers and other numbers relate to one another.

### Student 4

Student 4 shows that she can make sense of relationships between integers. She also shows that distance between endpoints or values in general pose some challenge. This is natural. For us, we want to provide diverse opportunities with all sorts of values to improve fluency with these distances. We might also arrange number lines with common endpoints to help students like her see connections. For example, we could use number lines of −20 and 20, −40 and 40, −200 and 200, as well as −400 and 400 to see how positional placements are similar and different as values become farther apart.

## Student Work 3

Use the endpoints to place the integer on the number line.

-5

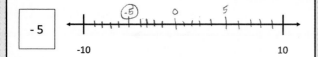

-10    10

- 12

-20    0

8

-10    10

-5

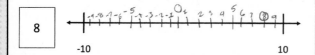

-30    30

## Student Work 4

Use the endpoints to place the integer on the number line.

- 5

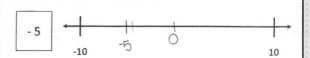

-10    10

- 12

-20    0

8

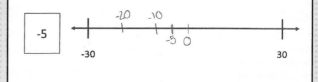

-10    10

-5

-30    30

Visit this book's companion website at **resources.corwin.com/ minethegap/6-8** for complete, downloadable versions of all tasks.

**TASK 10B:** Write the missing integers.

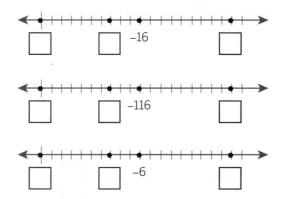

 **MINING TIP**

It is critical that our students work with number lines featuring varied, even unknown, endpoints. Doing this avoids misconceptions about locations and relationships on number lines.

*Students who work with number lines that feature the same endpoints may develop misconceptions about number relationships.* It provides a known and asks students to find missing integers, including endpoints. It's also designed to see if students recognize relationships between number lines as each features knowns and unknowns in the same positions. Students will likely count on or back from the known to find the missing values. Will our students find the missing values in the first number line and simply subtract 100 to find the values in the second?

**TASK 10C:** The • represents –25. Write the integer that each tick mark might represent.

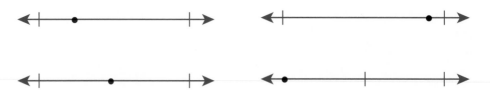

This task is a different take on task 10A. In it, students are given a specific location for an integer and have to reason about the endpoints. We should look for students who identify benchmarks. Do our students attempt to find zero so they can identify where –25 would be located? Or, are our students able to work with negative integers solely? This is most likely to appear in the two lower number lines. For example, they might write 0 as the right endpoint for the lower left number line or as the midpoint for the lower right number line. We should keep in mind that there are a range of possibilities for the lower left-hand model (0 and –50, –24 and –26, –20 and –30).

There is also a range of possibilities for the lower right-hand line, including tick marks of −24 and −23 (counting by ones) or −20 and −15 (counting by fives).

**TASK 10D:** Show −12 on a number line. Then, show −12 on a new number line with different endpoints. Tell how you knew where to place −12 on the second number line.

Task 10D is open-ended. It requires students to consider how −12 relates to other integers in two different situations. It is a good task for students to use independently or with partners before convening the class to discuss the range of possibilities. As we do this, we want to ask, How did you identify your endpoints? What else did you know about the relationships between the integers? Did you connect your thinking to other number situations (whole numbers, fractions, etc.)? Students who include zero in both number lines may show a reliance on zero for understanding negative integers. Others may overly rely on the midpoint and place −12 at the midpoint in both situations. Students may show that they need to count by intervals of 1, while others show that they can rely on larger intervals, benchmarks, or approximations.

NOTES

# BIG IDEA 11
## More Representing Integers on Number Lines

### TASK 11A

−30 is the midpoint for each number line, but each number line has different endpoints.

**What integer could the ? represent on each number line?**

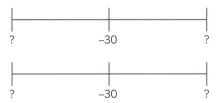

Tell how you found the endpoints for one of the number lines.

**MINING TIP**

Integer number lines that always feature zero as the midpoint may reinforce misconceptions about the relationship between integers.

### About the Task

Use of number lines is critical for developing understanding of integers and the relationships between integers. Yet, number lines can lead to misconceptions and misunderstanding. This may happen if we rely on the same endpoints each time we use number lines in our classrooms. It may also happen if *zero is always the midpoint of our number lines.* In this task, students have to consider how −30 relates to different integers on two different number lines.

## PAUSE AND REFLECT

• How does this task compare to tasks I've used?

• What might my students do in this task?

Visit this book's companion website at **resources.corwin.com/minethegap/6-8** for complete, downloadable versions of all tasks.

## Anticipating Student Responses

Student responses for this task might be quite interesting. We may find students who identify the right endpoint as 30 because of their ideas about positive and negative numbers. Others may place zero as the right endpoint. Some students may create tick marks between –30 and the endpoints and then count to find the endpoints. Others may reason about how –30 relates to other integers with many possibilities for endpoints including –31 and –29, –40 and –20, –50 and –10, or even –100 and 40.

NOTES

### Student 1

Student 1's work and writing show an understanding of the relationship between 15 and 30. He also shows that negative numbers are to the left and positive numbers are to the right. However, he doesn't show how these integers relate to one another in any way, including on a number line. His second number line shows that he does have some ideas about how he can manipulate numbers through their relationships, though that work is also inaccurate.

### Student 2

Student 2 shows understanding of how a midpoint relates a number with zero and another number in the first number line. In the second, he shows how a number can be halfway between two nonzero numbers. He shares that he is thinking of counting and that he counts on from intervals to find different points on the number line. It is also clear that he is still developing understanding of integers on number lines, as in both cases the absolute values increase from left to right as whole numbers would but do not for negative numbers.

## USING EVIDENCE

*What would we want to ask these students? What might we do next?*

### Student 1

Student 1 may be challenged by the midpoint being provided. It may be that his work with number lines has always provided endpoints and so he thinks about relationships between numbers in a narrow way. His use of 15 in the first number line is promising, as it shows he is thinking about "half" of a value, but it is flawed. We should begin by working with number lines that show midpoints and work to count on by any interval and then count back by the same interval. We should be sure to lift up how and why a number is half of another number or halfway between two specific numbers. We may first choose to use zero as one endpoint before working with more diverse endpoints.

### Student 2

Student 2 shows understanding of how numbers can relate to one another. He indicates how 30 is related to 0 and 60 as well as 15 and 45, though his use of the number line is reversed. Our work is to help him better understand how integers are plotted on a number line. We might have him compare a number line with values greater than zero with a number line showing their opposite values. We might also consider providing him with accurate placements of the numbers he selected and have him compare his work with the correct versions to see if he is able to make sense of the difference. We should work to practice and reinforce this understanding as it is critical for deep understanding of integers and the ability to compute with them.

## TASK 11A: −30 is the midpoint for each number line, but each number line has different endpoints. What integer could the ? represent on each number line?

### Student Work 1

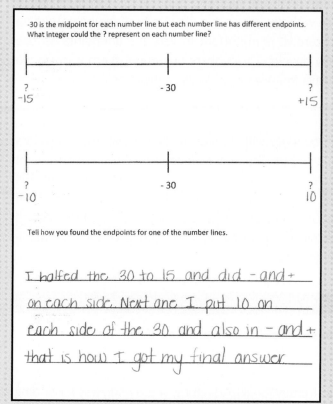

-30 is the midpoint for each number line but each number line has different endpoints. What integer could the ? represent on each number line?

?
−15

- 30

?
+15

?
−10

- 30

?
10

Tell how you found the endpoints for one of the number lines.

I halfed the 30 to 15 and did − and +
on each side. Next one I put 10 on
each side of the 30 and also in − and +
that is how I got my final answer

### Student Work 2

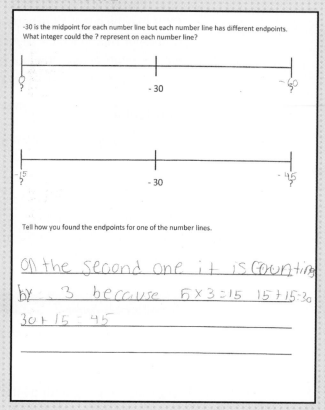

-30 is the midpoint for each number line but each number line has different endpoints. What integer could the ? represent on each number line?

0
?

- 30

−60
?

−15
?

- 30

−45
?

Tell how you found the endpoints for one of the number lines.

On the second one it is counting
by 3 because 5 × 3 = 15  15 + 15 = 30
30 + 15 = 45

### Student 3

Student 3's first number line is correct. He shows that he skip counts by multiples of 10 to find his endpoints. He doesn't replicate that strategy in the second number line. We should also take notice that both number lines use zero as one of the endpoints. This may be due to his skip counting in the first and duplicating it in the second. Or, it may be a lack of experience with diverse endpoints or a limited understanding of integer relationships beyond zero. It is important to vary the endpoints we use with number lines. Students may develop misconceptions about them if the same endpoints, such as zero, are always featured in tasks or instruction.

### Student 4

Student 4 seems to consider number lines with integers in the same way he considers number lines with whole numbers. But, Student 4 does show how we can count on and back by different intervals to establish endpoints for a given midpoint. His writing indicates that he doesn't fully understand the notion of "halfway" or "half of" as he notes a number is both in between and near. The latter is less precise and disconnected from the meaning of halfway as −330 and 270 are also viable endpoints, though neither is "near" −30.

## USING EVIDENCE

*What would we want to ask these students? What might we do next?*

### Student 3

We want to highlight Student 3's strategy for skip counting. It is a good way for finding numbers that are halfway between other numbers. As we know, it is imperative that we count on and back by the same interval to establish accurate endpoints. Student 3 does not do this in his second number line. We should also take note that the correct endpoint for the second number line is just above. This is a good opportunity to discuss how his tick marks above show equal distances from 30 when the second number line does not. Conversations like this as well as multiple opportunities to manipulate and discuss number relationships will help Student 3. We also must be sure to use endpoints that extend beyond zero.

### Student 4

Our most pressing action with Student 4 is to revisit representations of integers on number lines. He does not show that he understands that values decrease from left to right. As with Student 2, we might develop this understanding by comparing and contrasting number lines with various values, including integers. We should also be on the lookout for a student who seems to understand the concept when both positive and negative values are placed on a number line but is unable to do so when only negatives are placed on a number line. Students may show some understanding of left-to-right relationships with integers when both positive and negative values are present on a number line. However, they may not be able to transfer that understanding to number lines that only feature negative numbers.

**TASK 11A:** –30 is the midpoint for each number line, but each number line has different endpoints. What integer could the ? represent on each number line?

## Student Work 3

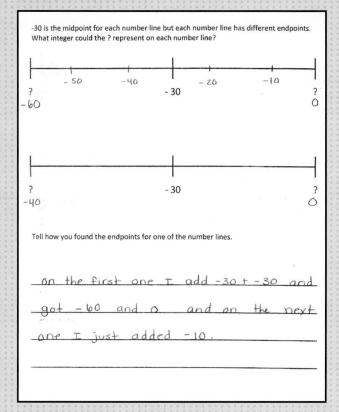

-30 is the midpoint for each number line but each number line has different endpoints. What integer could the ? represent on each number line?

Tell how you found the endpoints for one of the number lines.

on the first one I add -30 + -30 and got -60 and 0 and on the next one I just added -10.

## Student Work 4

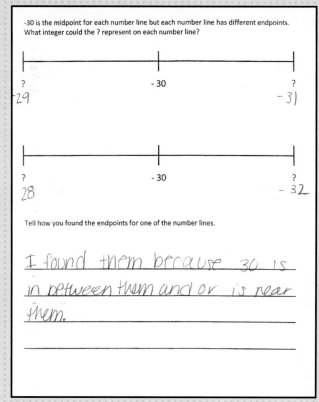

-30 is the midpoint for each number line but each number line has different endpoints. What integer could the ? represent on each number line?

Tell how you found the endpoints for one of the number lines.

I found them because 30 is in between them and or is near them.

## OTHER TASKS

- What will count as evidence of understanding?

- What misconceptions might you find?

- What will you do or how will you respond?

 Visit this book's companion website at **resources.corwin.com/ minethegap/6-8** for complete, downloadable versions of all tasks.

**TASK 11B:** Use the given midpoint to find the missing endpoints. Tell how you found the missing endpoints for one of the number lines.

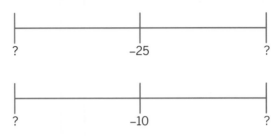

 **MINING TIP**

**Correct answers do not always indicate complete understanding. Asking students to represent and explain their thinking is typically more valuable than completing a large quantity of problems.**

Task 11B is a different take on how we can leverage midpoints to better understand our students' *thinking*. It features different values for each midpoint. *Our students' solutions will tell us a good bit about how they reason and generalize. Even correct answers may yield incomplete understanding.* For example, a student who counts by 1 in either direction (e.g., –26 and –24) on both number lines may reveal limited thinking about integers. A student who counts by a larger interval (5) but the same interval for both number lines may indicate limited thinking about number lines. In either case, it's critical that we talk with our students about why and how they completed this and all other tasks as it's possible that both of these students simply completed the task as quickly as possible.

**TASK 11C:** Both number lines have an endpoint of –15. The missing integers are different on each number line.

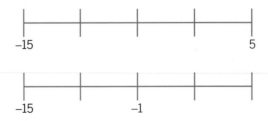

**Create a new number line that has different values than the two number lines above.**

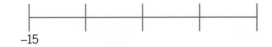

**Tell how you found the missing integers for one of the number lines.**

Task 11C investigates if students recognize that intervals on number lines are not set. Each number line is ticked, and it is essential that students understand relationships on ticked number lines before working with empty number lines. The intervals in this task are set (5 and 7). The second number line may be more challenging because the interval is 7 and the midpoint is negative. This number line is an important addition because it's possible that some students will successfully complete the first number line simply by recognizing the fives. The third line is open-ended. Will our students generalize that the tick marks can represent any interval? Or, will our students count on by 1? All of this tells us what we need to do next.

**TASK 11D:** **Chad says that both points represent the same integer. Matt says that can't be true because the number lines have different endpoints. Who do you agree with? Use pictures, numbers, or words to explain your thinking.**

Task 11D is a different take on reasoning about integer relationships with number lines. Students may agree with Chad that both points represent the same integer. These students indicate notions of fixed positions for values on number lines. These students likely believe that 50 or 12 is always the midpoint of a number line. They may not consider how a specific integer relates to others in flexible ways. Tasks like this are good opportunities for warm-ups or opening routines. When used this way, we can ask students to find missing values independently before bringing the class together to discuss solutions. It may be that we will need to inject other values or intervals into the discussion. With exposure and practice, our students will develop better understanding of integers and relationships while expanding their number sense beyond whole numbers.

NOTES

# BIG IDEA 12
## Comparing Integers

### TASK 12A

> Oscar says that −420 < −204. Do you agree with him? How can this be possible if 420 is greater than 204? Use pictures, numbers, or words to explain your thinking.
>
> Sam says that −900 + −10 + −2 > −912 because it has addition. Do you agree with him? Use pictures, numbers, or words to explain why you agree or disagree with Sam.

### About the Task

Comparing integers is a foundational concept. It is grounded in understanding of the meaning of place value and the relationship between integers. For some of our students, early misconceptions about comparing whole numbers may reappear. In this task, students are asked to address two of these misconceptions. The first examines digits and place value as well as the meaning of negative numbers in general. The second prompt asks students to compare an integer and the expanded form of the integer. It is a forerunner for comparing inequalities with integers and operations. However, computation with integers is not necessary for success with this task as it connects to ideas about decomposition of integers similar to that of task 9A.

### Anticipating Student Responses

Our students will agree or disagree for a host of reasons. For some, their ideas about positive integers may lead them to disagree that −420 < −204.

### PAUSE AND REFLECT

- How does this task compare to tasks I've used?

- What might my students do in this task?

Visit this book's companion website at **resources.corwin.com/minethegap/6-8** for complete, downloadable versions of all tasks.

Others may notice the same digits in both numbers, stating that the comparison can't be true. *Successful students will use a number line or a context to justify their reasoning.* Students may overlook that −900 + −10 + −2 is simply an expanded form representation of −912. Some may add the integers incorrectly or agree that addition always results in something greater.

 **MINING TIP**

Even correct justifications may indicate challenges. Students who often rely on the same model or context to describe a mathematical situation may be revealing a narrow or limited understanding of the concept. It may also be indicative of the models or contexts that are predominant during instruction.

NOTES

**MINING TIP**

Correct answers can occur for the wrong reasons or only work in certain cases. We should be on the lookout for reasoning that is mathematically accurate in all cases.

### Student 1

Student 1 notes that 420 is greater than 204, but because the integers are opposite, the comparison is opposite. *This reasoning works when comparing two negative integers, but it will be problematic when comparing a negative and a positive.* Her reasoning in her second response is flawed. She notes that addition creates larger quantities, but as we know, addition of negative numbers creates smaller quantities. She also misunderstands that the two values are equal.

### Student 2

Student 2 uses a number line to justify her comparison of −420 and −204. Her number line also shows a value greater than zero to reinforce her argument. Her second response indicates that she is able to compare values in some situations but not others.

## USING EVIDENCE

*What would we want to ask these students? What might we do next?*

### Student 1

We should investigate if Student 1's reasoning is more sophisticated than arguing that opposite comparisons happen with opposite values. We should ask her how or why a certain negative number would be less than a positive. It may be that she is muddling keep/change rules associated with operations with integers. We also want to develop understanding that comparisons can be justified by reasoning about number lines or similar representations. We also need to work with varied representations of values, including operations similar to the second prompt.

### Student 2

We can make use of Student 2's number line model to begin work developing understanding of the second prompt. It may be that she hasn't been exposed to addition with integers. It may also be attributed to unfinished learning surrounding decomposition of integers. We should work to help her with both and compare those results to other numbers. We should make use of varied representations but may first use a number line as she shows understanding of that model in her first response.

**TASK 12A:** Oscar says that −420 < −204. Do you agree with him? How can this be possible if 420 is greater than 204? Use pictures, numbers, or words to explain your thinking.

Sam says that −900 + −10 + −2 > −912 because it has addition. Do you agree with him? Use pictures, numbers, or words to explain why you agree or disagree with Sam.

## Student Work 1

Oscar says that -420 < -204. Do you agree with him? How can this be possible if 420 is greater than 204?

Use pictures, numbers, or words to explain your thinking.

I agree because 420 > 204 and this is negatives so it is opposite

Sam says that -900 + -10 + -2 > -912 because it has addition. Do you agree with him?

Use pictures, numbers, or words to explain why you agree or disagree with Sam.

I agree because addition makes it larger

## Student Work 2

Oscar says that -420 < -204. Do you agree with him? How can this be possible if 420 is greater than 204?

Use pictures, numbers, or words to explain your thinking.

I Agree because its a negtive number and in this situation -420 is lower

Sam says that -900 + -10 + -2 > -912 because it has addition. Do you agree with him?

Use pictures, numbers, or words to explain why you agree or disagree with Sam.

I agree because its a negtive number and there lower than zero

### Student 3

Student 3 cites the origin in both of her responses. The term aligns better with coordinate planes, though the reasoning in this work is sound. Essentially, she is establishing that values increase, or get bigger, as you move to the right of zero. At first, we might think that she is only applying the idea to positive integers. However, we can see that she connects it to values on both sides of zero and more broadly to any number. Her second response shows flawed understanding of adding with integers.

### Student 4

Similar to other students, Student 4 is able to justify her comparisons by relating integers to zero and by representing them on number lines. Also like other students, she has difficulty justifying why –900 + –10 + –2 isn't greater than –912. She discusses the idea of losing money as a context for comparison. She conveys that –900 + –10 + –2 is losing *more* money.

## USING EVIDENCE

*What would we want to ask these students? What might we do next?*

### Student 3

We can see that Student 3 has a foundation for comparing integers in the first prompt. It is also clear that she has misconceptions about adding with integers. Her idea about moving close to zero is reasonable for comparing values, however, her notion about what happens when adding negative numbers is problematic. She is working with the idea that adding moves numbers to the right on a number line. She connects this misconception of adding integers by incorrectly adding –900 + –10 and then –890 + –2. Like Student 2, we should begin to work with or revisit ideas about decomposition and operation with integers.

### Student 4

We can work with Student 4's ideas about losing more when adding negatives. Unlike the other students, she appears to understand that losing more or combining losses decreases the value. We should first investigate what she would determine the sum of the losses in the prompt to be. Is she able to find it to be –912 as well? Is her reasoning about adding negatives well described but fundamentally misunderstood? Is she unable to add negative numbers accurately? Is she able to use representations such as a number line to show addition with negative numbers? Each answer will outline a different path for us to take.

**TASK 12A:** Oscar says that −420 < −204. Do you agree with him? How can this be possible if 420 is greater than 204? Use pictures, numbers, or words to explain your thinking.

Sam says that −900 + −10 + −2 > −912 because it has addition. Do you agree with him? Use pictures, numbers, or words to explain why you agree or disagree with Sam.

## Student Work 3

Oscar says that -420 < -204. Do you agree with him? How can this be possible if 420 is greater than 204?

Use pictures, numbers, or words to explain your thinking.

...-10 -9 -8 -7 -6 -5 -4 -3 -2 -1 0 1 2 3 4 5 6 7 8 9 10 ...
← Getting bigger          Getting bigger →

O is the origin. tl is the small'st whole number. −1 is the bigest negative whole number.

Sam says that -900 + -10 + -2 > -912 because it has addition. Do you agree with him?

Use pictures, numbers, or words to explain why you agree or disagree with Sam.

−900 + −10 = −890 + −2 = −888

AS I said on the first problem O's the origin so I agree with Sam because when you add with negative nummbers the number gets closer to the origin, there for −888 is gratter than −912.

## Student Work 4

Oscar says that -420 < -204. Do you agree with him? How can this be possible if 420 is greater than 204?

Use pictures, numbers, or words to explain your thinking.

Because −204 is closer to O its greater then −420 in value. Numberline↓

-420 ——— -204 ——— 0 ——— 204 ——— 420

Sam says that -900 + -10 + -2 > -912 because it has addition. Do you agree with him?

Use pictures, numbers, or words to explain why you agree or disagree with Sam.

I disagree because when you lose money it says like −$750 meaning you lost money. Its a value below 0.

# OTHER TASKS

- What will count as evidence of understanding?
- What misconceptions might you find?
- What will you do or how will you respond?

 Visit this book's companion website at **resources.corwin.com/ minethegap/6-8** for complete, downloadable versions of all tasks.

**TASK 12B:** Use the number line to show how

- −58 compares to 45 and 32
- 19 compares to −22 and −10.
- −3 compares to −40 and 10.

 **MODIFYING THE TASK**

We could also ask students to write two comparisons as a result of their placements on each number line (e.g., −58 < 32 or −58 < 45). A nice extension for this task might be to have students add additional values to their number lines after placing the first three.

*In this task, students are prompted to use a number line to show how integers compare to other integers.* Students may make use of benchmarks to situate their integers. Placement on an open number line doesn't have to be exact but should be reasonable. Disproportionate placement may show that students don't fully understand the relationship between the values and are simply relying on the digits, place value, and/or the negative sign. Simply, it may show procedural understanding. An example of this would be a student who places the three integers on the number line with the middle value being the "midpoint" and the others being the endpoints (i.e., in the first prompt, −58 and 45 would be endpoints and 32 would be the midpoint).

**TASK 12C:** Write three numbers less than −62. Use pictures, numbers, or words to show that one of your numbers is less than −62.

Write three numbers greater than −19. Use pictures, numbers, or words to show that one of your numbers is greater than −19.

Open-ended tasks, like task 12C, are wonderful windows into student thinking. Open-ended tasks naturally provide opportunity for differentiation. Debriefing the work also provides opportunity for students to expand their perception of numbers and diverse thinking. Some of our students may simply write three consecutive numbers for each prompt. Though correct, it may be indicative of emerging understanding. Other students may skip-count by a certain amount, such as 10. In the first prompt, these students would find −72, −82, and −92 to be less than −62. Other students may write more random, accurate values. Of course, some of our students may show misunderstanding of the relationship between integers and that negative values are less as you move farther from zero.

**TASK 12D:** Order the following numbers: 50, −23, −19, 15, 32, −35, −8.

**Use what you know about numbers and their distance from zero to help explain how you ordered the numbers.**

Task 12D is a prime opportunity to expose our students' understanding of both integers and comparison. It requires them to make use of the understanding of integers and possibly number lines that they have developed during initial instruction. It is an opportunity for our students to find solutions and articulate their understanding. Doing both is much more powerful than oodles of mindless comparisons on a worksheet. Will they order the integers correctly? Will they use a model? Will they describe that they look for a negative sign and then consider place value? Can they communicate that the farther a positive number is from zero, the greater it is, or that the opposite is true for negative numbers? It will also be interesting to see if any students use other values beyond those provided in the prompt.

NOTES

# BIG IDEA 13
## Addition With Integers

### TASK 13A

> Jerry added −27 + 13 and got −14. He wrote −27 + 13 = −14.
>
> The next three times he added, his sum was less than −14.
>
> Write the numbers he might have added.
>
> Use pictures, numbers, or words to explain your thinking for one of the equations above.

### About the Task

Reasoning about addition with integers is critical for understanding the results of calculations. This impacts our students' ability to consider the reasonableness of their sums. Considering what happens when we combine positive and negative integers and the relationship between addends aids the development of computational fluency with integers. *In this task, students are provided with a true equation and asked to create three new equations based on a stated condition.* As with other tasks, discussing the wide range of results with the class can help develop generalizations and deeper understanding of the mathematics.

**MODIFYING THE TASK**

This task can be easily modified to feature different integers or operations.

### Anticipating Student Responses

The task is intended to determine if students can make sense of the relationship between addends as well as positive and negative numbers. Students might adjust one of the addends in each equation to be less than one of the original

## PAUSE AND REFLECT

- How does this task compare to tasks I've used?

- What might my students do in this task?

Visit this book's companion website at
**resources.corwin.com/minethegap/6-8**
for complete, downloadable versions of all tasks.

addends. Others might adjust both addends. Some students might adjust the addends correctly but find an incorrect sum, though the sum is less than the original. Others might adjust the addends incorrectly and find an incorrect sum that does satisfy the prompt. Some students might fully show a calculation to prove their new sum and how it compares to the original. Students with more advanced reasoning may simply explain how their adjusted addend(s) must result in a lesser sum.

NOTES

### Student 1

Student 1 replicates the digits in the prompt. He moves the negative sign to different locations in each line. His work may show a misconception related to fact families. It may also be connected to thoughts about multiplication and division with integers. His explanation below may be evidence of the latter.

### Student 2

Each of Student 2's expressions is accurate in relationship to the prompt. He justifies one of his sums by showing how he would add on a number line. Interestingly, Student 2 begins to approach the intent of the task by unchanging the first addend. Yet, he doesn't quite establish that the sum of −27 and another number will be less than −27 and 13 when that number is less than 13.

## USING EVIDENCE

*What would we want to ask these students? What might we do next?*

### Student 1

This task is intended to shed light on students' reasoning about how sums are influenced by changing addends. However, Student 1 uncovers a more significant misunderstanding. He is unable to reason about changing sums because he fundamentally misunderstands addition with integers. He may be a good example of what happens when students are prematurely taught rules in mathematics without deeper conceptual understanding. It is clear that we need to restart our work with addition of integers with Student 1. We want to be sure that in doing so we fully disconnect any ideas about the "rules" for adding integers. Instead, we should shift our focus to what is happening and why it is happening when integers are added. We should also be sure to use contexts and varied representations to develop this understanding.

### Student 2

Student 2 shows understanding that we can build on. He is able to find sums of integers. Our next work is to help him begin to reason about the results of his computations. We can do this in all sorts of ways. One approach is to consistently estimate or discuss what a sum or solution might be before finding the actual. After finding the actual, we can then compare results. This type of approach helps our students develop reasonableness of their answers.

## Student Work 1

Jerry added -27 + 13 and got -14. He wrote -27 + 13 = -14.

The next three times he added his sum was less than -14.

Write the numbers he might have added.

$$-27 \; + \; -13$$

$$27 \; + \; 13$$

$$27 \; + \; -13$$

Use pictures, numbers, or words to explain your thinking for **one** of the equations above.

negative Plus negat = negative

## Student Work 2

Jerry added -27 + 13 and got -14. He wrote -27 + 13 = -14.

The next three times he added his sum was less than -14.

Write the numbers he might have added.

$$-27 \; + \; 12$$

$$-27 \; + \; 5$$

$$-27 \; + \; 0$$

Use pictures, numbers, or words to explain your thinking for **one** of the equations above.

-27 + 5

-27  -26  -25  -23 -22

### Student 3

Like Student 2, Student 3 does not change the first addend in each of his expressions. We can leverage this to help him reason about how changing the second addend will affect the sum. His writing relates how adding a positive number to a negative number will impact the sum. Essentially, he has the ingredients for advancing his reasoning skills.

### Student 4

Student 4 shows quite a generalization for creating expressions that will satisfy the prompt. In two of the three expressions, he adds two negative numbers, each containing a negative that is quite less than the sum of the original equation. His third expression does add a negative and a positive, but he compensates by creating a considerably less negative number while leaving the second addend unchanged.

**USING EVIDENCE**

*What would we want to ask these students? What might we do next?*

### Student 3

As noted, Student 3 has ingredients for reasoning about sums of negative and positive numbers. We might approach his development similar to Student 2. They might both benefit from discussion with one another. We might also be sure to work with situations in which similar equations are recorded in a fashion that helps them see relationships. *For example, we might record –27 + 13 = –14, –27 + 12 = –15, and –27 + 11 = –16 on top of one another.* We can then ask students to describe what remains the same and what changes with each equation. We can ask them to discuss the magnitude of each change. We might also ask them to predict what the next equation might be. Examining these sorts of patterns within computations may be new to our students and so we should remain patient and resilient by sticking with the activity for several days.

**MINING TIP**

Our recording of expressions and equations must be intentional so that students can see patterns and relationships between them. Many of us were not taught to record mathematics this way on chalk or white board.

### Student 4

Student 4 shows a different type of reasoning that is useful when reasoning about computation. However, it may not be intentional, so it is up to us to investigate further. Essentially, he shows in his second expression that two addends will generate a smaller sum if they are both less than two other addends. His third prompt builds on the idea of unchanged addends. Like the others, Student 4 shows ideas about numbers that need nurturing. We should be sure to offer consistent opportunities to play with expressions or equations and to discuss our results so that students develop the type of reasoning that goes well beyond mechanical computation.

## TASK 13A: Jerry added −27 + 13 and got −14. He wrote −27 + 13 = −14. The next three times he added, his sum was less than −14. Write the numbers he might have added. Use pictures, numbers, or words to explain your thinking for one of the equations you wrote.

### Student Work 3

Jerry added -27 + 13 and got -14. He wrote -27 + 13 = -14.

The next three times he added his sum was less than -14.

Write the numbers he might have added.

-27 + 5

-27 + 10

-27 + 8

Use pictures, numbers, or words to explain your thinking for **one** of the equations above.

-27+5=-22,because if you add a positive number to a negative number it gets the negative number closer to zero.

### Student Work 4

Jerry added -27 + 13 and got -14. He wrote -27 + 13 = -14.

The next three times he added his sum was less than -14.

Write the numbers he might have added.

-57 + -1

-176 + -35

-296 + 13

Use pictures, numbers, or words to explain your thinking for **one** of the equations above.

I know sence -296 is much lower then -27 the answer should be lower sence your adding the same number.

# OTHER TASKS

- What will count as evidence of understanding?
- What misconceptions might you find?
- What will you do or how will you respond?

 Visit this book's companion website at **resources.corwin.com/ minethegap/6-8** for complete, downloadable versions of all tasks.

**TASK 13B:** Add each. Use a number line to show how you added.

$$35 + 45 \qquad -37 + -36 \qquad 28 + -42 \qquad -15 + 61$$

Task 13B provides insight into how students combine integers. It helps us see if they decompose integer addends in ways similar to whole numbers. The first prompt is intentional as it provides "baseline" information into their approaches to adding in general. Some may add 35 + 40 + 5. Others may add 45 + 30 + 5, while others may think of 30 + 40 + 10. There are a host of ways to manipulate the various expressions. Students may decompose by place value or use friendlies. Students can use either endpoint as a starting point for the computation or work within endpoints of a specific range (e.g., 0 and 100 for 35 + 45). We should look for students who move to the right in each of the four expressions. We should also look for students who add the expressions procedurally to find the sums and then draw on the number line to match the equations.

**TASK 13C:** Break apart one or both numbers to make them easier to add. Write the sum.

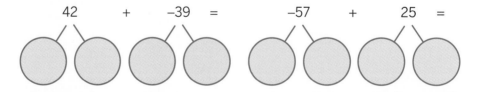

42     +     −39     =          −57     +     25     =

**MODIFYING THE TASK**

This task features a positive and negative integer addend. It can be modified to feature two positive or two negative integers. In time, we may consider adding a third addend.

*As with the other tasks, task 13C is intended to promote number sense and computational fluency.* It makes use of number bonds to represent how an integer addend might be decomposed to add to another integer. We want to look for students to decompose in rational ways. In the first prompt, they may think of 42 as 39 and 3. In doing so, they can add −39 + 39 + 3 making the computation considerably easier. In the second, students might think of −57 as −25 and −32. The addition then becomes −25 + 25 + −32. Of course, there are all sorts of ways to decompose integers and other numbers. In this task, we want to look to see if our students decompose to improve the efficiency of calculation.

**TASK 13D:** Use the integer chart to add −38 + 41, −29 + 27, −12 + −22, and −32 + 59.

| −40 | −39 | −38 | −37 | −36 | −35 | −34 | −33 | −32 | −31 |
|-----|-----|-----|-----|-----|-----|-----|-----|-----|-----|
| −30 | −29 | −28 | −27 | −26 | −25 | −24 | −23 | −22 | −21 |
| −20 | −19 | −18 | −17 | −16 | −15 | −14 | −13 | −12 | −11 |
| −10 | −9 | −8 | −7 | −6 | −5 | −4 | −3 | −2 | −1 |
| 0 | 1 | 2 | 3 | 4 | 5 | 6 | 7 | 8 | 9 |
| 10 | 11 | 12 | 13 | 14 | 15 | 16 | 17 | 18 | 19 |
| 20 | 21 | 22 | 23 | 24 | 25 | 26 | 27 | 28 | 29 |
| 30 | 31 | 32 | 33 | 34 | 35 | 36 | 37 | 38 | 39 |
| 40 | 41 | 42 | 43 | 44 | 45 | 46 | 47 | 48 | 49 |
| 50 | 51 | 52 | 53 | 54 | 55 | 56 | 57 | 58 | 59 |

This task makes use of an integer chart introduced earlier in this chapter (task 9C). Adding with integer chips may work well for many of our students, but the practicality of the addends is limited. Number lines may also work well. Adding on an integer chart helps students understand how they can apply whole number addition strategies to integers. Using the chart, we may see them decompose an addend and count on by tens and then ones. In time, we should see them count on by multiples of tens and ones. The integer chart also helps students see how combining positive and negative numbers can be done by working through zero. For example, in the expression −38 + 41, we might decompose the 41 into 38 and 3. Then, we would combine or add −38 and 38 to find 0. 0 and 3 more equals 3. So, −38 + 41 = 3.

NOTES

# BIG IDEA 14
## Subtraction With Integers

### TASK 14A

Kaleb found the difference of 4 different pairs of integers to be −7. What might the 4 different pairs of integers be? Use models, numbers, or words to justify your solutions.

## About the Task

Subtraction can be thought of as taking away or finding the difference between two values. In this task, students solve an open-ended problem with subtraction using either of these ideas about subtraction to justify their results. Open-ended problems such as task 14A provide opportunities for students to demonstrate understanding, practice computation, and develop reasoning. These sorts of problems offer a variety of possible solutions. These problems also can provoke productive struggle as they don't offer a specific, clear solution path. Students can be exposed to different strategies and reasoning as they examine the work of classmates.

**MINING TIP**

Calculations are efficient. We don't want to force students to represent every problem once they demonstrate understanding. However, calculations alone do not always indicate full understanding.

## Anticipating Student Responses

This task is likely to uncover students who *add* two integers to find a sum of −7 instead of a difference. At first glance, we may think they simply misunderstood the prompt, but it is just as likely that they don't understand how to subtract with integers. Success will occur for a variety of reasons as well. *Some students may rely on models to justify their solutions. Others*

### PAUSE AND REFLECT

- How does this task compare to tasks I've used?
- What might my students do in this task?

Visit this book's companion website at **resources.corwin.com/minethegap/6-8** for complete, downloadable versions of all tasks.

*may rely on calculations. Still others may leverage their understanding of patterns and operations after justifying one solution.* For example, they might establish that 5 – 12 = –7. These students then may add 1 to both numbers (6 – 13) or subtract 1 (4 – 11) repeatedly to create the four equations.

**MINING TIP**

We must be careful to avoid admonishing students who make use of patterns. Though it may seem like a shortcut, it is evidence of noteworthy reasoning. It would be wise to have students share their reasoning with classmates before posing the all-important question, "Will it always work?"

NOTES

### Student 1

Student 1 offers a few different ideas that should grab our attention. Her subtraction expressions do not have a difference of −7. Her other two equations do not represent subtraction. Her first model does represent her expression, which is addition. Her second model shows misunderstanding of how zero pairs are made. Her representations for addition are accurate.

### Student 2

Like Student 1, Student 2 creates two equations that make use of addition. The idea of using integer chip representations is encouraging, but it is important to note that the first two examples do not show subtraction. They do represent how zero pairs could be used to show the amount of −7. Student 2's writing also conveys a limited perspective of addition and/or subtraction with integers.

## USING EVIDENCE

*What would we want to ask these students? What might we do next?*

### Student 1

We might think that Student 1 overlooked that the prompt asked for a difference of −7. Even so, there is other evidence that she needs more time developing understanding of subtraction with integers. We can build off of her understanding of integer chips and addition. We should also be sure to reinforce the meaning of subtraction. It is clear that her ideas about taking away or finding the difference between two integers is not well understood. This may even contribute to her use of addition expressions as she wasn't sure that her first two expressions were correct so she was compelled to show what she *could* do.

### Student 2

As noted with Student 1, we need to develop understanding of subtraction with integers. Using integer chips or similar models can help students as they work with this concept. It is also possible that these students (and others in this set) think about subtraction as "take away." Though accurate, the thinking is limited. As comfort with subtraction grows, we should begin to introduce the idea that subtraction also describes the difference between two values. When doing this, we can use a number line to find the difference between two points on it. As with other concepts, learning the mathematics concepts is enhanced when used in context.

## Student Work 1

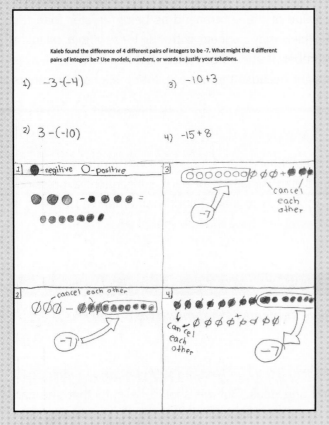

## Student Work 2

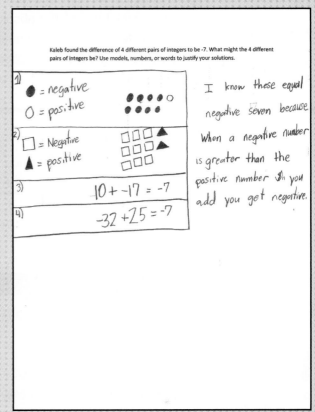

### Student 3

Student 3 offers four expressions that are equal to –7. She uses integer chips to justify each expression. In each, she draws the subtrahend and takes away the minuend. We can also see that each subtrahend is less than the minuend. Or, we might think of the absolute value of the subtrahend as being greater than the absolute value of the minuend, which may connect with her ideas about taking a smaller number from a larger number. It should be noted that her first equation is incorrect, but it is likely an oversight because the representation aligns with others that are accurate.

### Student 4

Student 4's ideas are telling. She established an expression with a difference of –7. She then adjusted the subtrahend and minuends of each subsequent expression. Making use of patterns is a desirable way to learn about mathematics. However, it should be noted that she doesn't establish how she knows that –6 – 1 = –7.

*What would we want to ask these students? What might we do next?*

**MINING TIP**

Open-ended tasks such as this are good opportunities to see what our students understand. When using them, we should look for patterns within students' answers. Student 3's pattern (using all negatives) may signal incomplete understanding.

### Student 3

We must ask ourselves, *"Does Student 3 fully understand subtraction with integers?" Her work seems to indicate that she does.* But we should be sure that she can successfully subtract integers in which the minuend is less than the subtrahend such as –4 – (–11). We should also investigate if she can subtract integers that don't share like signs. It may be that she found these expressions to be easiest to represent accurately. It may also be that this is her only understanding of subtraction with integers.

### Student 4

Student 4's use of patterns is admirable. We want to encourage her and others to look for and make use of patterns. We should be sure that she understands why –6 – 1 = –7. Knowing our students helps us determine if she does, in fact, know the difference without needing a justification. We should also investigate if Student 4 can successfully subtract other integers that have a difference of –7. We should look to see if she builds from a known equation or is able to subtract without using a pattern. For example, we might ask her if the difference of –31 and –24 is –7. Will she be able to subtract, or will she have to build on from her pattern or something similar?

## Student Work 3

Kaleb found the difference of 4 different pairs of integers to be -7. What might the 4 different pairs of integers be? Use models, numbers, or words to justify your solutions.

* −11−4=−7
  - I know that is correct because :
  ⊝⊝⊝⊝ ⊝ ⊝ ⊝ ⊝⊝⊝ ✱negative chips(⊝)

* −8−(−1)=−7
  - I know that is correct because :
  ⊝⊝⊝⊝⊝⊝⊝ ⊝

* −9−(−2)=−7
  - I know that is correct because :
  ✱negative chips(⊝)
  ⊝⊝⊝⊝ ⊝⊝⊝⊝

* −21−(−14)=−7
  - I know that is correct because :
  ⊝⊝⊝⊝⊝⊝⊝⊝⊝⊝⊝⊝⊝⊝⊝⊝⊝⊝⊝⊝⊝

## Student Work 4

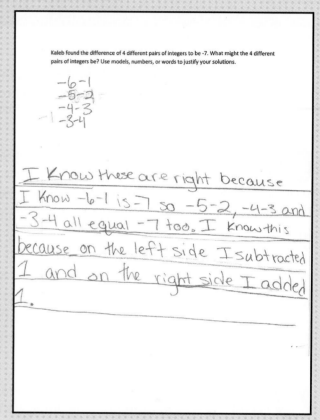

Kaleb found the difference of 4 different pairs of integers to be -7. What might the 4 different pairs of integers be? Use models, numbers, or words to justify your solutions.

−6−1
−5−2
−4−3
−3−4

I Know these are right because
I Know −6−1 is −7 so −5−2, −4−3 and
−3−4 all equal −7 too. I Know this
because on the left side I subtracted
1 and on the right side I added
1.

## OTHER TASKS

- What will count as evidence of understanding?
- What misconceptions might you find?
- What will you do or how will you respond?

 Visit this book's companion website at **resources.corwin.com/minethegap/6-8** for complete, downloadable versions of all tasks.

**TASK 14B:** Solve each expression. Then, use number lines, integer chips, or another model to justify your thinking.

$$-8 - 7 = \qquad 13 - (-6) = \qquad -5 - (-12) =$$

**MINING HAZARD**

Students should be able to represent a concept in various ways. Consistent examples of the same representation may indicate incomplete understanding or representational bias.

*As students represent subtraction of integers, they indicate their understanding of the concept.* The representation they use offers insight into how they understand the concept. It may even show which models they prefer. It may also show the models we feature during instruction. The models they select may be linked to the specific situation. It may be easier to prove 13 − (−6) by thinking of taking away −6. However, −8 − 7 may be easier to model as the difference between the two points on a number line. We might modify the task to ask students to represent each with two different models. It's also important to recognize that a context for completing each is also relevant. For example, they might think of −8 − 7 as the temperature at −8 degrees that then falls 7 more degrees.

**MINING HAZARD**

Correct responses may not indicate complex or full understanding. We may be fooled to think that a certain response was a choice of taking an easy path to completing the task. It may be the only path, which requires us to dig deeper into their understanding.

**TASK 14C:** The difference of two different sets of integers is −14. What are the possible integers? Use numbers, pictures, or words to justify your answers.

An open-ended task such as 14C helps us in different ways. It will enable us to see if students recognize that different integers can have the same difference (or sum, for that matter). An inability to do so may be rooted in misconceptions about whole numbers and/or operations. This task also offers insight into the situations or integers students prefer. We may have a student who thinks of −14 as −13 − 1 and −12 − 2. We may find that others can think about it more complexly. They may offer −27 − (−13) or 14 − 28. *Though both instances are correct, students who represent it as 1 less and 2 less may not yet understand the concept, operation, or computation as fully.*

**TASK 14D:** We know that −75 − (−50) = −25.

**Use it to tell if each expression below is greater or less than −75 − (−50) = −25.**

$$-75 - (-45) \qquad 75 - 50 \qquad 0 - 30 \qquad -88 - (-63)$$

Reasoning can be more efficient than calculating. At the very least, reasoning helps us determine if the results of our calculations are reasonable! This task can be completed in a variety of ways. Some students will simply calculate each expression and compare it to the results of the prompt. It is intended to determine if students can use known information $-75 - (-50) = -25$ to evaluate the results of other expressions. Students who reason about the situations may offer all sorts of ideas about how they compared differences to $-25$. Some predominant examples may be that $-75 - (-45)$ is greater because they "see" a difference of 30. $75 - 50$ should be a quick mental computation to easily compare, though some may see it as equal because the digits are the same. The expression $0 - 30$ is less because it, too, is an easy calculation. Reasoning for $-88 - (-63)$ may be more sophisticated as some students may note that both numbers are 15 less so the difference remains the same. *This task can be easily used as an opening number routine in which we ask students to participate without pencil and paper.* We may need to begin with whole numbers before shifting to fractions, decimals, and integers.

**MODIFYING THE TASK**

**This task can be easily modified to compare other expressions or to provoke other misconceptions. The intent should remain the same. Simply, we want to see if our students mindlessly compute or if they reason about the numbers being computed.**

NOTES

# BIG IDEA 15
# Multiplication With Integers

### TASK 15A

> Compare the two multiplication facts on each line. Circle the fact that has the greater product.
>
> | | |
> |---|---|
> | 6 × −6 | −6 × −4 |
> | 9 × 1 | −5 × 8 |
> | −10 × −2 | −6 × −10 |
> | 8 × 9 | −5 × −7 |

**MODIFYING THE TASK**

Intentionality is critical with tasks like these. We may design it so that specific situations are compared throughout the task. For example, it might be a collection of positive × positive compared with positive × negative to reinforce the "rule" of multiplying integers.

## About the Task

Multiplication with integers can be much more than recalling that a negative times a negative is a positive. Granted, that foundational understanding is necessary so that our students can apply understanding and reasoning to new situations and tasks such as this. In this task, students are asked to compare the products of two multiplication expressions. We can use tasks like these in all sorts of ways. We may use them as an opening routine or activity. In doing so, we may have students compare expressions independently before exchanging ideas with a partner and then discussing as a class. *We may even ask students to engage in these tasks without paper and pencil so that reasoning and mental mathematics can be promoted.*

## PAUSE AND REFLECT

- How does this task compare to tasks I've used?

- What might my students do in this task?

Visit this book's companion website at **resources.corwin.com/minethegap/6-8** for complete, downloadable versions of all tasks.

## Anticipating Student Responses

Our students will approach this task in all sorts of ways. We will have students who compute each product so that they can compare the expressions. Some may compute mentally. Some will use models to find their products. Those who use mental computations may then use a model to justify their comparisons. Some students will make use of limited or no computation to compare the expressions. For example, they may note that $9 \times 1$ is greater than $-5 \times 8$ because it yields a positive product. A similar thought may be applied to $6 \times -6$ and $-6 \times -4$. These students may compare the expressions in the third line by considering the number of groups or that the expressions share a factor.

NOTES

### Student 1

Student 1 makes three incorrect choices. His work is interesting in that he appears to have originally selected −6 × −4 but clearly changed his mind. His writing below reinforces the logic we would expect for correct comparisons.

### Student 2

Three of Student 2's responses are correct. He originally identified −6 × −10 as greater than −10 × −2 but changed his mind. His explanation reveals considerably large gaps. *Fortunately, he decided to explain his incorrect selection.* His writing reveals a misunderstanding about multiplying integers (−10 × −2 ≠ −20). He also states that "with negative numbers the least amount is the greatest," revealing a misconception about comparison as well.

## USING EVIDENCE

*What would we want to ask these students? What might we do next?*

### Student 1

Student 1 provides a challenge. Two of his selections are correct and two are not. His reasoning appears valid and applicable to the task. He is a good example of when more digging is needed. It may be noteworthy that both of his incorrect comparisons contain an expression with one negative number, whereas the correct comparisons have expressions with two negative factors or two positive factors. It may be that he has unfinished learning relative to products of integers. One way to dig deeper is to ask him to explain his other comparisons. Another may be to provide new comparisons that only use his possible misunderstanding to see if he reproduces the errors.

### Student 2

Student 2's correct responses may mislead us to believe he is ready to move forward. This isn't so. He shows that he doesn't fully understand multiplication with integers, and he may have flaws with comparison of integers. Using tasks that combine or connect concepts is useful for checking in on previously learned concepts as this task connects with comparing integers. But when tasks uncover possible misconceptions for varied concepts, we must isolate specific challenges. Our next step with Student 2 is to investigate his understanding of comparison and multiplication of integers separately so we can determine what to work with specifically.

## Student Work 1

Compare the two multiplication facts on each line.
Circle the fact that has the greater product.

(6 ×-6)　　　　　-6 ×-4

9 ×1　　　　　(-5 ×8)

-10 ×-2　　　　　(-6 ×-10)

(8 × 9)　　　　　-5 ×-7

Choose one of the lines you compared. Use models, numbers, or words to tell how you compared the products.

I compared the product by looking at the numbers, 'cause -6×4 equals -24, and 6×-6 equals -36, but positives are greater than negative.

## Student Work 2

Compare the two multiplication facts on each line.
Circle the fact that has the greater product.

6 ×-6　　　　　(-6 ×-4)

(9 ×1)　　　　　-5 ×8

(-10 ×-2)　　　　　-6 ×-10

(8 × 9)　　　　　-5 ×-7

Choose one of the lines you compared. Use models, numbers, or words to tell how you compared the products.

-10×2 is greater than -6×-10 because -10×-2=-20 and -6×-16=-60 -20 is greater because with negative numbers the least amount is the greatest because it's closer to 0 or zero.

### Student 3

Student 3 successfully compares each equation. He uses a model to justify his products of 9 × 1 and −5 × 8. Curiously, his representation of −5 × 8 doesn't show an accurate array, though there are 40 negatives. Additionally, his legend connects with the representations, but what does it tell us about the models? Does he clearly state that circles are positives and squares are negatives? Is it clear that −5 × 8 = −40?

### Student 4

Each of Student 4's comparisons is correct. His explanation cites the rules for multiplication of integers. His writing indicates that he didn't need to compare the values because one product is positive and the other is negative.

**USING EVIDENCE**

*What would we want to ask these students? What might we do next?*

### Student 3

Student 3's representations may cause us to pause and ask questions. We should be sure that he can also represent other situations with multiplication of integers. For example, it would be interesting to see how he represents −10 × −2. Our instructional next steps will be clear if he is unable to do so. Assuming he can and does understand multiplication with integers, we should then shift our work to reasoning about it. Citing the rules for multiplication with integers is useful for efficient comparison once understanding is in place. We can connect these ideas with notions about comparing expressions such as expressions that share one factor (−10 × −2 and −6 × −10).

### Student 4

Student 4 shows understanding of how to multiply with integers. His explanation may not be as clear as we would like as he doesn't give the exact products of both expressions. But it may be as simple as he doesn't need to consider products when one is negative and the other is positive. We can move Student 4's reasoning forward by adding more factors to an expression or by focusing his reasoning on efficiency rather than exact values. Like Student 3, we can discuss comparison of expressions that share a factor. *We might discuss how we can think about expressions in which both factors are greater than the other (8 × 9 and −5 × −7).*

**MINING TIP**

Discussing reasoning strategies with students who can quickly and accurately recall basic facts can be difficult because they simply know the products. We can adjust expressions to feature more complex numbers so that discussion focuses on ideas rather than exact calculations. For example, we could adjust 8 × 9 to 83 × 97 and −5 × −7 to −51 × −72 to discuss that both factors in the first expression are greater so their product will be greater.

**TASK 15A:** Compare the two multiplication facts on each line. Circle the fact that has the greater product.

## Student Work 3

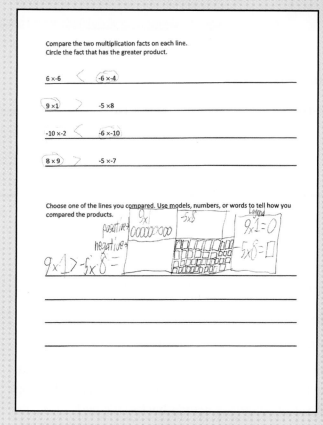

## Student Work 4

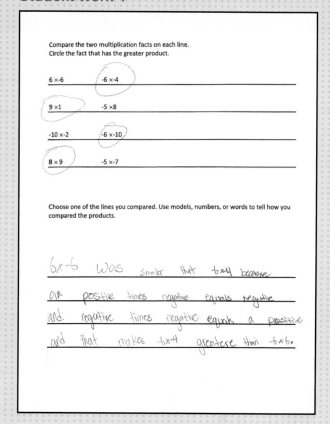

OTHER TASKS

- What will count as evidence of understanding?
- What misconceptions might you find?
- What will you do or how will you respond?

 Visit this book's companion website at **resources.corwin.com/minethegap/6-8** for complete, downloadable versions of all tasks.

**TASK 15B:** Review the equations.

| | | | | | | | | | |
|---|---|---|---|---|---|---|---|---|---|
| 4 | × | 4 | = | 16 | 4 | × | −4 | = | −16 |
| 3 | × | 4 | = | 12 | 3 | × | −4 | = | −12 |
| 2 | × | 4 | = | 8 | 2 | × | −4 | = | −8 |
| 1 | × | 4 | = | 4 | 1 | × | −4 | = | −4 |
| 0 | × | 4 | = | 0 | 0 | × | −4 | = | 0 |
| −1 | × | 4 | = | −4 | −1 | × | −4 | = | 4 |
| −2 | × | 4 | = | −8 | −2 | × | −4 | = | 8 |
| −3 | × | 4 | = | −12 | −3 | × | −4 | = | 12 |
| −4 | × | 4 | = | −16 | −4 | × | −4 | = | 16 |

**What patterns do you notice in the equations?**

**How might these patterns help you when you multiply integers?**

Mathematics is a science of patterns. Patterns help us generalize and make shortcuts. They can also be used to explain results. *In this task, students may notice that the first factor in each set of equations counts backwards by 1.* Students might notice that the product counts backwards by one group of the second factor (4 and −4). They might also notice that positive × positive and negative × negative yield positive products, whereas positive × negative factors yield a negative product. This and additional work with patterns of integer factors can further support the understanding of how to discern if a product is positive or negative.

**MINING TIP**

Intentional recording of mathematical ideas is critical for discovering patterns. As with this task, factors and products are aligned in a specific sequence so that students can look for and confirm patterns.

**TASK 15C:** The number line shows expressions and their products plotted on a number line. Write the missing multiplication expressions. How might this representation help you think about multiplying with integers?

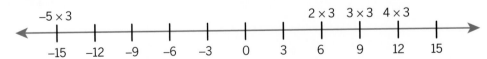

This task is a similar take on the idea of leveraging patterns and models to develop understanding of concepts. Students should notice that the points on the number line represent products of 3. In other words, it is counting groups of 3. To the right, we can see 2, 3, and 4 groups of 3. To the left, we only see −5 groups of 3. Students can make use of the skip counts or the recognizable basic facts. We should look for students who reverse the factors. These expressions will yield the same product but mean something slightly different. Students may note that the number line or pattern of expressions can help them remember if a product is negative or positive. They might also note that it can help them find other products when a specific expression is "known" or that they can use the commutative property to find products. Students with inaccurate expressions, especially positive expressions, may show fundamental multiplication misunderstandings.

## TASK 15D: Oscar knows −3 × 4 = −12.

**How can he use that to help him solve −6 × 4 and −6 × 8?**

**Use pictures, numbers, or words to explain your thinking.**

We can leverage understanding and properties of multiplication to develop efficient mental strategies. The same is true for multiplication with integers. In this task, students are presented with a known and asked to use it to find the products of other expressions. Computations may be used by some students, but doing so doesn't realize the intent of the prompt. We should look for students who recognize that the product of −6 × 4 is twice that of −3 × 4 because −6 is twice −3. Students may then use that idea to find the product of −6 × 8. Some students may not see the connection between all three and instead consider the product of −6 × 8 to be 4 times −12 because −6 is twice −3 and 8 is twice as much as 4. *Tasks that make use of related expressions are prime opportunities for discussion and reasoning.*

**MODIFYING THE TASK**

**Our students can help us create tasks of related expressions or number strings to use in whole class situations. First, we expose them to these sorts of tasks and then ask them to create their own examples of related expressions or number strings. We can then select the most interesting examples to use with the entire class.**

NOTES

# BIG IDEA 16
## Division With Integers

### TASK 16A

The number line shows expressions and their quotients plotted on a number line. Write the missing division expressions.

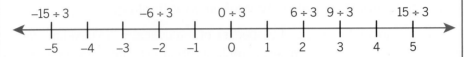

Tell how you found the missing expressions.

How might this representation help you think about dividing with integers?

### About the Task

It is critical that our students make meaning of the mathematics through various models. We can support this by varying the representations or models we use during instruction. Students should be able to transfer the ideas or concepts to different representations if they are understood deeply. In this task, students connect quotients on a number line with related expressions. It is intended to complement work with division of integers that features integer chips or contextual representations. The task could also be leveraged to examine relationships between expressions. These related expressions could be recorded on top of one another. For example, $-20 \div 4$ could be recorded on top of $-15 \div 3$ because both have the same quotient.

## PAUSE AND REFLECT

- How does this task compare to tasks I've used?
- What might my students do in this task?

 Visit this book's companion website at **resources.corwin.com/minethegap/6-8** for complete, downloadable versions of all tasks.

## Anticipating Student Responses

This task provides opportunity to see if students understand the meaning of division as well as the +/− result of division with integers. Some students may write any expression with a quotient equal to the location on the number line. Doing so misses the point that we are thinking about groups of 3. There will be students who demonstrate this understanding but change the sign of the divisor (3) to −3, overlooking the increase in the dividend as we move from left to right on the number line. Students who successfully find the missing expressions may be unable to communicate why or how they found them beyond recall of basic facts. Better developed understanding may be evident as students refer to the number of groups of 3 in an amount (dividend). Similarly, students may note that each time they are finding how many groups of 3 are in varied amounts that increase from left to right on the number line. Students may note that this helps them remember if the result of division will be positive or negative.

NOTES

## Student 1

Student 1 shows that she can find quotients of integers. Her expressions do not use divisors of 3 as provided in the task, yet they are connected. She makes use of $-8 \div 2$ and $8 \div 2$ for quotients of $-4$ and $4$, respectively. Her response to the second prompt does not offer any new ideas.

## Student 2

Unlike Student 1, Student 2 does find a pattern on the number line. She notes that each is dividing by 3 and records the expressions we expected. Her second response notes that multiplication can be used to help divide. The task hopes to uncover that there are other ways to prove that a negative divided by a positive is a negative. However, her thinking here is relevant and useful.

### *What would we want to ask these students? What might we do next?*

## Student 1

Student 1's ideas are encouraging, though she doesn't approach the task in the ways that we would like. She shows a relationship between expressions with similar factors. She tells how she found the quotients but doesn't connect the divisors within the tasks. We may first direct her to that specifically. We may also provide all of the expressions for her and students like her to highlight and discuss the pattern of counting by 3s. We should focus our work here and with different divisors before shifting the conversation to discuss how this helps us think about division with integers.

## Student 2

Student 2 provides expected expressions, a description of the pattern, and an idea about how the representation is useful. There is nothing wrong with her ideas, but she didn't draw a conclusion that the task is intended to find. It may be that the second prompt should be adjusted to specifically ask if the representation can help us draw a conclusion about dividing a negative number. It's also possible that regardless of the prompt, she doesn't see the relationship between the divisors, dividends, and quotients. Open questions or explanations might not yield the insight that we desire in an activity or prompt. However, they do provide insight into a student's perspective. We can adjust them to be precise. But in doing so, we might lose access to other ideas that we can use to advance student thinking. There is value in her idea about connecting multiplication and division. It can be used to leverage models of multiplication to connect and understand the results of division with integers. It can also be used to later reinforce recall of the rules of multiplication and division with integers because the two are explicitly connected.

**TASK 16A:** The number line shows expressions and their quotients plotted on a number line. Write the missing division expressions.

Tell how you found the missing expressions.

How might this representation help you think about dividing with integers?

## Student Work 1

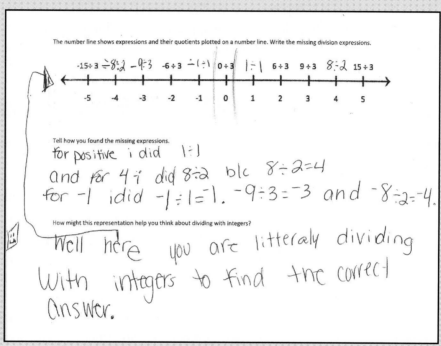

The number line shows expressions and their quotients plotted on a number line. Write the missing division expressions.

-15÷3  -8÷2  -9÷3   -6÷3  -(-1)  0÷3  1÷1  6÷3  9÷3  8÷2  15÷3

-5  -4  -3  -2  -1  0  1  2  3  4  5

Tell how you found the missing expressions.

for positive i did 1÷1
and for 4 i did 8÷2 b/c 8÷2=4
for -1 idid -1÷1=-1. -9÷3=-3 and -8÷2=-4.

How might this representation help you think about dividing with integers?

Well here you are litteraly dividing with integers to find the correct answer.

## Student Work 2

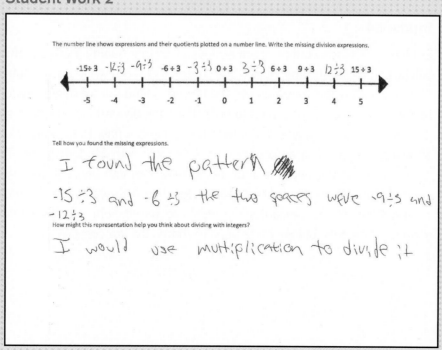

The number line shows expressions and their quotients plotted on a number line. Write the missing division expressions.

-15÷3  -12÷3  -9÷3   -6÷3  -3÷3  0÷3  3÷3  6÷3  9÷3  12÷3  15÷3

-5  -4  -3  -2  -1  0  1  2  3  4  5

Tell how you found the missing expressions.

I found the pattern

-15÷3 and -6÷3 the two spaces weve -9÷3 and -12÷3

How might this representation help you think about dividing with integers?

I would use multiplication to divide it

## Student 3

Student 3 creates one expression. She notes a rule on the chart but doesn't offer what it is. Her response to the second prompt notes that quotients can be either negative or positive but doesn't note *when* they are negative or positive.

## Student 4

Each of Student 4's expressions on the number line is accurate. Her responses aren't robust, but they do share her approach to finding each new expression. We should take note that her second response is an observation rather than an insight.

### *What would we want to ask these students? What might we do next?*

## Student 3

As we know, there are different reasons why our students' work is incomplete. We must avoid making assumptions about incomplete work. Student 3 is a good example. Before making decisions about what she has offered, we should ask her if she can tell us anything about the missing expressions. We might simply ask, "Do you think there could be an expression for these other numbers?" It's also possible that incomplete work is simple misunderstanding or simply a different perspective. She offers $-9 \div 3$. Interestingly, that is the only positive expression that is missing a related negative expression. We might need to direct her to create expressions for $-4$ and $4$ as well as $-1$ and $1$. It's a challenge to move forward without understanding *why* the work is incomplete.

## Student 4

Student 4 reminds us that quantity doesn't dictate quality. Even so, efficient should still be complete. She shares that she subtracted 3. But what did she subtract 3 from? She may not be able to remember the term dividend but we might expect to see "the first number." We might expect to note that each dividend is 3 less, or she subtracts 3 from each as she moves to the left on the number line. Her second statement isn't an insight; rather, it is an observation. Our work with Student 4 and those like her is to develop better depth to their responses. This can be done through group and class discussion, student dictations for teachers to record, and eventually student writing itself. We can encourage students to be efficient with sentence complexity and use of models, but we must be sure that ideas are complete.

**TASK 16A:** The number line shows expressions and their quotients plotted on a number line. Write the missing division expressions.

Tell how you found the missing expressions.

How might this representation help you think about dividing with integers?

## Student Work 3

The number line shows expressions and their quotients plotted on a number line. Write the missing division expressions.

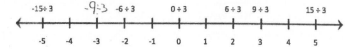

Tell how you found the missing expressions.

I look on the chart to find it out to rule out the one that was missing

How might this representation help you think about dividing with integers?

It Shows how a divison problem can have a negative or positive answer

## Student Work 4

The number line shows expressions and their quotients plotted on a number line. Write the missing division expressions.

Tell how you found the missing expressions.

I Just Subtracted 3

How might this representation help you think about dividing with integers?

Some were negtive.

## OTHER TASKS

- What will count as evidence of understanding?
- What misconceptions might you find?
- What will you do or how will you respond?

 Visit this book's companion website at **resources.corwin.com/ minethegap/6-8** for complete, downloadable versions of all tasks.

**TASK 16B:** Show −28 ÷ −4 in two different ways.

Tasks don't have to be overly complicated to deeply examine student understanding. This task simply asks for two different representations of −28 ÷ −4. Some students may show that there are 7 groups of −4 in −28 using integer chips. Others may show something similar with a number line model or with a contextual example such as the number of $4 withdrawals needed for a total withdrawal of $28. −28 ÷ −4 = 7 is a representation, but it doesn't necessarily show deep understanding of the mathematics. Simply stating that a negative divided by a negative is a positive is also accurate, but without a better justification, the idea may not be fully understood.

**TASK 16C:** Emma knows 60 ÷ −6 = −10.

- How can she use it to find 54 ÷ −6?
- How can she use it to find 48 ÷ −6?
- How can she use it to find 72 ÷ −6?

Once our students demonstrate why the results of quotients are positive or negative, we can begin to develop their fluency with integer computation. In this task, students make use of a known equation to find other quotients. These expressions feature related basic facts, and so some students may recall and apply a rule to determine the result. This line of thought is fine but may be indicative of procedural thinking, or it may be challenging to extend it to more varied situations such as those that don't include basic facts. Other students may share that 54 ÷ −6 has one less group of 6 than 60 so there is one less group of −6, resulting in a quotient of −9.

**TASK 16D:** Ruth wrote these division facts.

| | |
|---|---|
| −50 ÷ 10 = −5 | −50 ÷ 5 = −10 |
| −40 ÷ 10 = −4 | −40 ÷ 5 = −8 |
| −30 ÷ 10 = −3 | −30 ÷ 5 = −6 |
| −20 ÷ 10 = −2 | −20 ÷ 5 = −4 |

**What patterns do you notice in the division facts?**

**So how can knowing −70 ÷ 10 = −7 help you solve −70 ÷ 5?**

Organizing expressions and equations intentionally can help our students recognize relationships and patterns within them. Our students might note that the left column adds 10 each time and that the divisor remains 10 so the quotient increases by 1. They might note that dividing by 5 is half of dividing by 10, so the quotients of integers and 10 will be double that of the same integers and 5. We could extend the pattern to positive dividends (e.g., –20, –10, 0, 10, 20, . . .). Extending these patterns and discussing them can help our students understand *why* a negative divided by a positive results in a negative.

NOTES

# RATIO, PROPORTION, AND PERCENT

- Big Idea 17: Representing Ratios

  Ratios can be represented in different ways. But ratios can also represent different ideas, including part-to-part and part-to-whole relationships.

- Big Idea 18: Equivalent Ratios

  Equivalent ratios model the same relationship between two quantities. We can use a variety of representations to model this equivalency.

- Big Idea 19: Unit Rates

  Finding and applying unit rates allows us to have a base of comparison between two scenarios. Unit rates also produce a scale factor for finding other equivalent ratios.

- Big Idea 20: Using Ratios to Solve Problems

  Ratio reasoning provides multiple strategies for solving real-world problems.

- Big Idea 21: Reasoning With Percents

  Tasks involving percent build on understanding of ratio reasoning, because percents are values compared to a whole of 100.

- Big Idea 22: Unit Rate as Slope

  Understanding unit rate as slope leads to understanding of linear functions. We can represent this relationship in different ways to determine the constant of proportionality.

# BIG IDEA 17
## Representing Ratios

### TASK 17A

**Which of the following represent a ratio of 2:3?**

**Explain your reasoning.**

### About the Task

Representing ratios helps our students understand and distinguish between part-to-whole and part-to-part relationships. *Deep understanding of ratio means that our students can recognize a specific ratio in different contexts, situations, or representations.* In this task, students are presented with a ratio and asked to identify different representations of this ratio. *The task determines whether students are considering the meaning of a ratio and whether they can identify nontraditional representations of the same ratio.*

### Anticipating Student Responses

Some students will be quick to identify D as a model of 2:3, noting that there are 2 shaded parts and 3 parts in the whole. Others, with similar thoughts, will also note that A is an example comparing the unshaded parts to the whole as well. Students with more developed understanding of ratio will identify these as well as B, as it shows 2 shaded parts to 3 unshaded parts. These students may overlook E, though it, too, shows a 2:3 relationship between unshaded and shaded parts.

 **MODIFYING THE TASK**

This task could be extended by asking students to offer new examples of the ratio 2:3.

**MINING HAZARD**

Overreliance on shaded pieces that identify a part of a whole may lead students to think that fractions or ratios only identify shaded parts to parts in a whole.

## PAUSE AND REFLECT

- **How does this task compare to tasks I've used?**

- **What might my students do in this task?**

 Visit this book's companion website at
resources.corwin.com/minethegap/6-8
for complete, downloadable versions of all tasks.

### Student 1

Student 1's response indicates that he understands a ratio to represent a part-to-whole relationship. He analyzed each ratio in the context of the number of shaded regions to the total number of regions. Student 1 also made a computational error writing an equivalent ratio for 6:10, which prevented him from identifying choice E as a correct representation. Student 1 has not considered the representations that model a part-to-part relationship.

### Student 2

Like Student 1, Student 2 is also focusing solely on a part-to-whole relationship. However, Student 2 also recognizes that the shaded region does not have to be represented by one of the ratio quantities. He accurately selects choice A because the diagram represents two unshaded regions to three total regions. By not selecting choice E, Student 2 is unable to consider representations of equivalent ratios. Student 2 has also not considered any of the part-to-part relationships.

*What would we want to ask these students? What might we do next?*

### Student 1

Student 1 will need exposure to ratios that represent part-to-part relationships. Examples that include different quantities in different contexts, such as fruits to vegetables, are one way for students to distinguish between part-to-whole and part-to-part relationships. Then, we can have Student 1 write as many ratios as possible for one given scenario or diagram. This will enable the student to think beyond one type of representation.

### Student 2

Unlike Student 1, Student 2 is able to flexibly see some part-to-whole relationships. We should be sure to acknowledge his understanding. However, like Student 1, Student 2 should also work with part-to-part relationships in context. We can introduce a scenario in which he must represent part-to-part relationships. As understanding develops, we can have Student 2 revisit this and similar tasks with his new perspective.

## Student Work 1

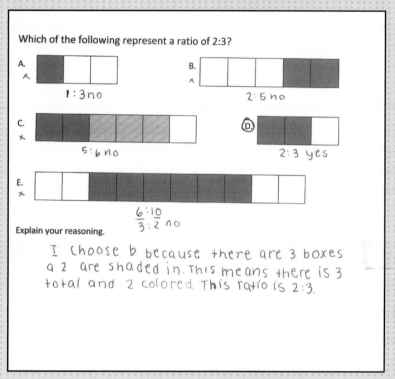

**Which of the following represent a ratio of 2:3?**

A.
✗
1:3 no

B.
✗
2:5 no

C.
✗
5:6 no

(D.)
2:3 yes

E.
✗

$\frac{6:10}{3:2}$ no

**Explain your reasoning.**

I choose b because there are 3 boxes a 2 are shaded in. This means there is 3 total and 2 colored. This ratio is 2:3.

## Student Work 2

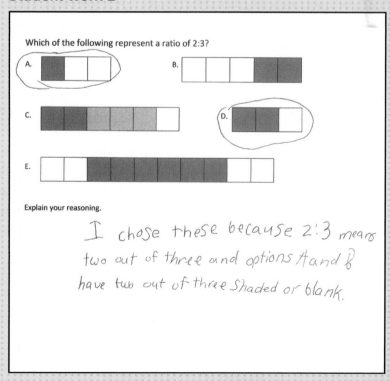

**Which of the following represent a ratio of 2:3?**

A.

B.

C.

D.

E.

**Explain your reasoning.**

I chose these because 2:3 means two out of three and options A and B have two out of three shaded or blank.

## Student 3

Student 3 has selected two representations for part-to-part relationships. In both cases, the student has identified a ratio to represent shaded and unshaded regions. For choice B, Student 3 uses the construct of number of shaded boxes to number of unshaded boxes. For choice E, the student was able to see a ratio of 4 unshaded boxes to 6 shaded boxes and recognize that this would be equivalent to the ratio of 2:3. The student shows flexibility with ordering of the quantities within the ratio. Student 3, however, is unable to identify any of the part-to-whole relationships.

## Student 4

Student 4 has selected some representations to model part-to-part relationships (choices B and C) and one representation to model a part-to-whole relationship (choice D). He accurately describes why each of his selections modeled a relationship of 2:3. Even so, he is unable to flexibly apply this reasoning to the other choices.

*What would we want to ask these students? What might we do next?*

### Student 3

Student 3 offers ideas to acknowledge and build on. He is flexible with representations for part-to-part relationships, including representations for equivalent ratios. Articulating this type of relationship will be important as he transitions to part-to-whole relationships. We can introduce Student 3 to situations in which he must identify parts of a whole and represent using a ratio. Tasks with these ratios in context can be used before transitioning to tape diagrams or similar representations. We can then ask him to compare examples of part-to-part ratios as part-to-whole ratios.

### Student 4

Since Student 4 has shown the ability to recognize both part-to-part and part-to-whole relationships, it is important to highlight these differences as we debrief with him and others like him. It will be important to understand why Student 4 did not select choices A or E in order to gain a better understanding of his thought process. We should ask Student 4 to explain why the other choices are not examples and then encourage him to consider nonshaded regions to total regions or nonshaded regions to shaded regions. His misconceptions may be the result of the representations he experiences during instruction.

### Student Work 3

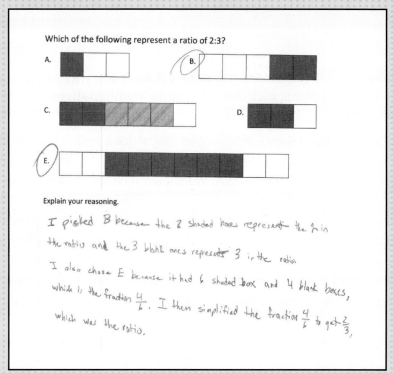

Which of the following represent a ratio of 2:3?

A.

B.

C.

D.

E.

Explain your reasoning.

I picked B because the 2 shaded boxes represent the 2 in the ratio and the 3 blank ones represents 3 in the ratio.
I also chose E because it had 6 shaded box and 4 blank boxes, which is the fraction $\frac{4}{6}$. I then simplified the fraction $\frac{4}{6}$ to get $\frac{2}{3}$, which was the ratio.

### Student Work 4

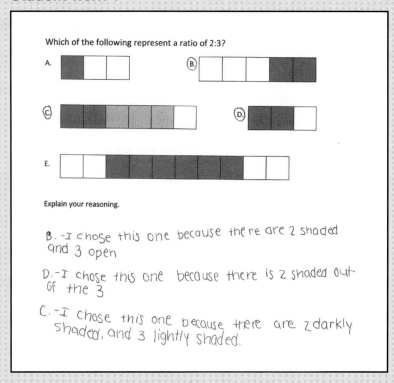

Which of the following represent a ratio of 2:3?

A.

B.

C.

D.

E.

Explain your reasoning.

B. - I chose this one because there are 2 shaded and 3 open

D. - I chose this one because there is 2 shaded out of the 3

C. - I chose this one because there are 2 darkly shaded, and 3 lightly shaded.

## OTHER TASKS

- What will count as evidence of understanding?
- What misconceptions might you find?
- What will you do or how will you respond?

 Visit this book's companion website at **resources.corwin.com/minethegap/6-8** for complete, downloadable versions of all tasks.

**TASK 17B:** Represent the ratio 3:4 as many ways as you can. You may use tape diagrams, pictures, diagrams, manipulatives, and fractions to represent.

Multiple representations of a mathematical idea help students develop a deep understanding of that idea. This task requires students to generate various representations for the ratio 3:4. We should look for students who create multiple representations of the same idea. For example, they might show a part-to-part relationship in varied ways. Even doing so, they don't show a complete understanding of 3:4. Proficient students will show a combination of part-to-whole and part-to-part relationships. They may use varied partitioned models (rectangles or something similar), sets of objects (circles to triangles), or contexts (baseballs to basketballs) to represent these ideas. We may even see equivalent examples or scale factors of 3:4, such as 6:8 or 9:12.

**TASK 17C:** Part A. Write a ratio that compares the prime to composite numbers below.

**100, 2, 3, 8, 12, 16, 20, 7, 50, 80**

Part B. Write a ratio that compares all shapes to quadrilaterals.

**MINING HAZARD**

Students who understand concepts may represent them incorrectly as symbols. It is critical that we connect representations and symbols during instruction and assessment.

**Explain how you determined the ratio for Part A or Part B.**

*This task focuses on representing both part-to-part and part-to-whole relationships.* It also highlights the notion that ratios can compare any set of ideas. Students may incorrectly interchange the order of the ratio. This misconception should be addressed early to avoid further error as students begin to explore proportional relationships. Proficient students will recognize or generate equivalent ratios. For example, in Part B, students may generate the ratio 6:4 or 3:2. This task is also a good example of how we might integrate different mathematics concepts.

**TASK 17D:** Maya plays on a soccer team. The table below shows how her team performed this year.

| Number of Games Played | Number of Wins | Number of Losses | Number of Home Games Won | Number of Away Games Won |
|:---:|:---:|:---:|:---:|:---:|
| 18 | 12 | 6 | 10 | 2 |

**Use the team data to write two ratios. Describe what each ratio represents.**

It is important that our students identify ratios of things that are being compared. We can ask students to create ratios for various situations. It is also important that our students work with ratios in real-world contexts. This task does just that. Students have to acquire and process the information provided in the table and then select two ratios to represent. They then must describe what the ratio means in the context of the problem. We should look for students who identify two different examples of part-to-part or part-to-whole ratios. This may be evidence of limited understanding. Students may also begin to rewrite ratios in simplest terms, such as 1:3 representing number of losses to number of games played. We should be careful to avoid overemphasizing simplest terms before students fully understand the meaning of ratio. Moreover, we should be careful to dismiss ratios not represented in simplest form as incomplete understanding.

NOTES

# BIG IDEA 18
## Equivalent Ratios

### TASK 18A

> Jenny is sure that 10:12 is equivalent to 25:30. Do you agree with Jenny?
>
> Use pictures, numbers, or words to justify your thinking.

### About the Task

**MODIFYING THE TASK**

**We can modify the numbers used in the ratio relative to our students' understanding and number sense development.**

Ideas about equivalent ratio extend well beyond the ability to calculate equivalency. Representations support students' understanding of the concept of equivalent ratios. *The open-ended nature of this problem allows the student to select his or her own representation and interpretation of each ratio.* The temptation to calculate may be limited due to the nature of the numbers in the ratios. This is especially noteworthy because there may be problems or tasks with numbers that enable students to find convenient "solutions," possibly without deep understanding. We should also note that this task prompts the students to agree or disagree rather than correct flawed ideas. Our students then need to verify their thinking.

**MINING HAZARD**

**Pictorial representations are appropriate ways to show relationships between two quantities, but using them to verify equivalence can be difficult as numbers or relationships become more complex.**

### Anticipating Student Responses

Students will typically represent the ratios using tape diagrams or double number line models, as these are the clearest way to show equivalence. *Some students may illustrate these ratios with pictorial representations. Some students may rely on contexts to make sense of ratios. Yet in this task, these*

## PAUSE AND REFLECT

- How does this task compare to tasks I've used?
- What might my students do in this task?

Visit this book's companion website at **resources.corwin.com/minethegap/6-8** for complete, downloadable versions of all tasks.

*students may then have difficulty justifying why 10 boys to 12 girls is equivalent to 25 boys to 30 girls.* For this type of task, students often represent the ratios as part-to-whole models, but some may choose to show a part-to-part relationship. Displaying a part-to-part relationship may be more challenging for students to verify equivalence. It may also be that some of our "number crunchers" may find that 25 is $2\frac{1}{2}$ times greater than 10 and then look for the same relationship between 30 and 12 to "prove" the two ratios are equivalent.

NOTES

### Student 1

Student 1 incorrectly determines that the two ratios are not equivalent. Her reasoning is that "25:30 can't be rounded to 10:12." This student has not shown evidence of using multiplicative strategies to generate other equivalent fractions. Unfortunately, she didn't attempt to represent the problem in different ways.

### Student 2

Student 2 has demonstrated that she understands that equivalent ratios have multiplicative relationships. Her thinking is flawed because she is only considering a scale factor of 2. For the first ratio, she multiplies both quantities by 2, and in the second ratio, she divides both quantities by 2. When the two resulting ratios are not the same, Student 2 incorrectly determines that the ratios are not equivalent. It appears that Student 2 was trying to use another representation, like a number line or counting strategy, but changed her thinking.

**USING EVIDENCE**

*What would we want to ask these students? What might we do next?*

### Student 1

We would first want to ask Student 1 to clarify what she means by "rounded to." This may uncover a considerable misconception about what equivalent ratios are. Next, we should ask her to share other strategies she might use to determine whether two ratios are equivalent. We can ask her to verify her thinking using one of these other strategies. If she is unable to generate an alternative strategy, we might first ask her to rewrite the ratio as fractions and then determine whether the fractions are equivalent. We may also need to revisit other strategies to help her extend her thinking beyond calculation.

### Student 2

Since Student 2 has shown some understanding of multiplying by a scale factor, it is important to build on this understanding. We would first want to have her summarize what she did and ask why she chose a scale factor of 2. We can ask her if the conclusion would be the same if she chose a different scale factor. We should provide her time to try different scale factors in order to see the relationship between the two ratios. Student 2 is a good example of a student who has misconceptions about ratio that develop from her number sense. Her work is a good reminder as to why we must vary the scale factor as students seem to show understanding with whole number scale factors.

**TASK 18A:** Jenny is sure that 10:12 is equivalent to 25:30. Do you agree with Jenny?

Use pictures, numbers, or words to justify your thinking.

## Student Work 1

Jenny is sure that 10:12 is equivalent to 25:30. Do you agree with Jenny?

Use pictures, numbers, or words to justify your thinking.

No, I don't agree with Jenny, because 25:30 can't be rounded to 10:12.

## Student Work 2

Jenny is sure that 10:12 is equivalent to 25:30. Do you agree with Jenny?

Use pictures, numbers, or words to justify your thinking.

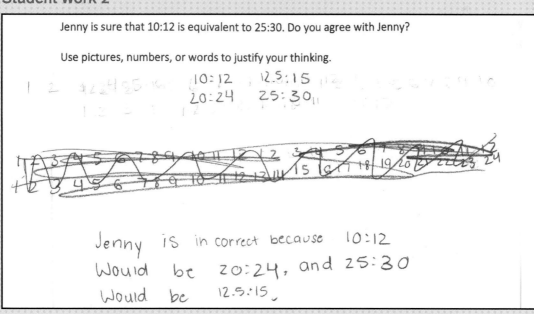

10:12    12.5:15
20:24    25:30

Jenny is in correct because 10:12 Would be 20:24, and 25:30 Would be 12.5:15.

### Student 3

Student 3 chose to divide quantities and compare quotients as a means to determine whether the ratios are equivalent. To compare two ratios, it is viable to divide one quantity within the ratio by the other quantity. We can also divide corresponding quantities between ratios. Her disagreement is the result of a miscalculation as she writes that $30 \div 12 = 5$ instead of 2.5.

### Student 4

Student 4 has a valid strategy for determining whether the two ratios are equivalent. She finds an equivalent ratio that we might consider to be simplest terms. She then finds if that new ratio is equivalent to the second ratio. Her justification shows that she knows how to find the solution, but not necessarily why this works.

### USING EVIDENCE

*What would we want to ask these students? What might we do next?*

### Student 3

We can build on Student 3's understanding that division may be used as a strategy to determine whether ratios are equivalent. We would want to ask the student to explain why she chose to divide 30 by 12 and 25 by 10, then ask how these values are related. We might also consider making tools available so that students can verify their calculations. We might investigate if she can apply this to finding if more than two ratios are equivalent. Highlighting her approach through class discussion would be wise as it can expose students to a different approach that they may be able to use as well.

### Student 4

Student 4 has a valid strategy for determining whether the two ratios are equivalent. Essentially, she is showing that if two different ratios are equivalent to a third ratio, then the two are equivalent. But her understanding seems to be more procedural than conceptual. Subsequent work with her should focus on monitoring her conceptual understanding of the task. We should ask the student why this strategy works or why this solution makes sense. This will help us gauge how this student understands the relationship between ratios.

**TASK 18A:** Jenny is sure that 10:12 is equivalent to 25:30. Do you agree with Jenny?

Use pictures, numbers, or words to justify your thinking.

## Student Work 3

Jenny is sure that 10:12 is equivalent to 25:30. Do you agree with Jenny?

Use pictures, numbers, or words to justify your thinking.

$30 \div 12 = 5$  not equal
$25 \div 10 = 2.5$

example:

12:30

10:50

$30 \div 12 = 5$  equal
$50 \div 10 = 5$

## Student Work 4

Jenny is sure that 10:12 is equivalent to 25:30. Do you agree with Jenny?

Use pictures, numbers, or words to justify your thinking.

$$\frac{10 \div 2}{12 \div 2} = \frac{5 \times 5}{6 \times 5} = \frac{25}{30}$$

Yes, they are equivelent, because $10 \div 2 = 5 \times 5 = 25$.
$12 \div 2 = 6 \times 5 = 30$.

You divide the number by 2, then multiply the divided numbers by 5, to determine if the ratios are equivelent or not.

## OTHER TASKS

- What will count as evidence of understanding?

- What misconceptions might you find?

- What will you do or how will you respond?

 Visit this book's companion website at **resources.corwin.com/ minethegap/6-8** for complete, downloadable versions of all tasks.

**TASK 18B:** **Which of the following ratios are equivalent to the ratio 3:4? For one of the equivalent ratios, justify your reasoning. Create a new example of a ratio that is equivalent to 3:4.**

$\frac{6}{8}$ 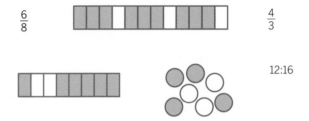 $\frac{4}{3}$

12:16

**MINING HAZARD**

**Identifying examples of equivalent ratios is not as cognitively demanding as creating examples of equivalent ratios. We should be sure to include both when determining our students' understanding.**

This task requires students to demonstrate understanding of equivalent ratios using a variety of representations. Students need to recognize equivalent fractions, part-to-whole relationships, and part-to-part relationships. Some students may only select responses that represent part-to-whole relationships, excluding the set of circles. Students may also consider the 12-section rectangle as a part-to-part relationship rather than the part-to-whole representation of 9:12. *Creating a new example also helps us determine how well our students understand equivalent ratios.* A picture of 6:8 (that has already been established as equivalent) is a different representation of 6:8 rather than a new example of an equivalent ratio such as 12:16.

**TASK 18C:** **Write three ratios that are equivalent to the ratio $\frac{12}{18}$. How do you know these ratios are equivalent?**

Often, students confront ratios in simplest terms when prompted to determine equivalency. This may cause them to think that they can find equivalence by multiplying. This task allows students to work in varied directions. Common strategies may include halving or doubling the quantities, simplifying the fraction, or multiplying by a common scale factor. Misconceptions typically occur when students do not recognize that ratio equivalence is centered around the multiplicative relationship between the quantities. We should be sure to look for other equivalence misconceptions resulting from addition or subtraction rather than multiplication or division.

**TASK 18D:** The sixth-grade class at Northern Middle School consists of 80 boys and 100 girls. The principal wants to plan a school event and decides to survey a sample of the class. If the samples are to be representative, make a recommendation to the principal about how many sixth-grade boys and girls to survey.

Ratios are useful for all sorts of situations and problems in our everyday lives. This task provides students with an opportunity to apply understanding of equivalent ratios to a real-world problem. *Students have the opportunity to select an appropriate sample size that could be used to represent the population of the school.* Students should generate samples that are equivalent to 80:100, but are significantly smaller numbers, such as 8:10. Students may find an equivalent ratio to 80:100, but have difficulty translating their solution in the context of the problem. This task would be a good opportunity for students to work with partners or in small groups to find their recommendations. We can then facilitate a whole group discussion about the possible solutions, noting similarities and differences. During this conversation, we should lift up the ideas of equivalency as well as practicality.

**CONTENT CONNECTION**

While this task is assessing understanding of ratio equivalence, it is also closely tied to Statistics and Probability concepts of sampling. Make sure students have an understanding of the vocabulary and key concepts prior to administering this task.

NOTES

NOTES

# BIG IDEA 19
## Unit Rates

### TASK 19A

> Sara ran 3 miles in 18 minutes. Jessica ran 5 miles in
> 28 minutes. Who is the fastest runner? Explain your reasoning.

### About the Task

Unit rates are an important concept that allows for quick comparison of two rates. Unit rates help us determine better values when buying goods or services. They also help us compare productivity and proficiency in sports. Simply, they are a tool for reasoning, analysis, and problem solving. This task is an example of how unit rates can be leveraged to compare the speed of different runners even though they run different distances in different amounts of time. Students must be able to explain who is the fastest runner. *To do this, students must generate one equivalent unit in order to compare the runners' speeds.*

**MODIFYING THE TASK**

We can extend the task by asking students to create a new time and distance for a third runner who is slower or faster than either or both runners in the prompt.

### Anticipating Student Responses

Students first must choose whether they want to compare based on miles or minutes. Students may try to compare how long it takes each runner to run one mile. If students choose to compare based on time, students may try to compare how many miles each runner runs in one minute. Some students may choose to not create unit rates, but rather compare based on a common multiple of one of the units, such as how long it takes each runner to run 15 miles. Any of these strategies are viable. We may also find that some students use calculations to make comparisons, while others use representations and calculations to find their solutions.

## PAUSE AND REFLECT

- How does this task compare to tasks I've used?
- What might my students do in this task?

 Visit this book's companion website at
**resources.corwin.com/minethegap/6-8**
for complete, downloadable versions of all tasks.

### Students 1 and 2

Students 1 and 2 both decided to find a unit rate based on the time it takes to run 1 mile. While the strategy is the same, there are sharp contrasts between the two students' work. Student 1 first begins by creating the ratio 3:5, but then decides this is not a valid strategy. He then creates two ratio tables for each runner and finds the time based on one mile. Student 1 offers no explanation for his work and no response to the question, "Who is the fastest runner?"

Student 2 applies the same strategy, but is also able to articulate some of his thinking. He has accurately labeled the ratio table to show the comparison of miles to minutes. Student 2 is able to accurately identify that Jessica is the fastest runner. Student 2's explanation only states that "Jessica runs faster because 5.6 is less than 6." This explanation does not account for the reason why the lower number represents a faster runner.

*What would we want to ask these students? What might we do next?*

### Student 1

We should first ask Student 1 to explain the ratio tables he has created. We might ask, "What do these quantities represent?" or "How did you decide to show the values in this way?" Understanding what he created leads us to ask him how this strategy might determine a unit rate. We should ask Student 1 to explain what each of these unit rates means in the context of this problem. We should be sure to listen for understanding of how the unit rate connects to the concept of minutes to run 1 mile. As we ask him questions about his thinking and representations, we should be careful to avoid funneling him to the ideas we want him to have.

### Student 2

Student 2 has a valid strategy and simply needs to provide some clarification of his reasoning. We may want to investigate how he established that Jessica runs 1 mile in 5.6 minutes as that may not be an easy mental computation. We also want him to explain how he knows that Jessica is the fastest runner. This connection is important because real-world contexts, problems, and results do not always align with an exact number. For example, this task could be adjusted to ask who is the slower runner or runs farther in 50 minutes. In each example, we think of the unit rate in slightly different ways to answer the question.

## Student Work 1

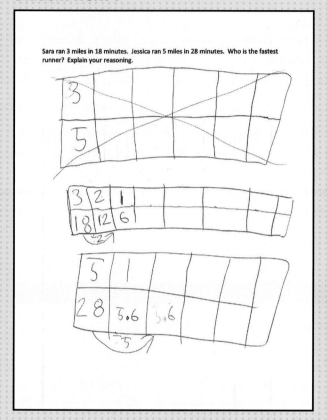

Sara ran 3 miles in 18 minutes. Jessica ran 5 miles in 28 minutes. Who is the fastest runner? Explain your reasoning.

## Student Work 2

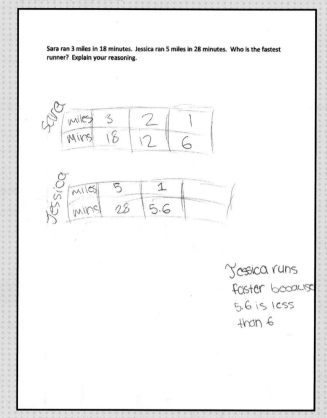

Sara ran 3 miles in 18 minutes. Jessica ran 5 miles in 28 minutes. Who is the fastest runner? Explain your reasoning.

Jessica runs faster because 5.6 is less than 6

### Students 3 and 4

Students 3 and 4 applied the same strategy to determine a solution. Both students chose to find a common number of miles run to compare. In this case, both students made a comparison based on running 15 miles. Their work and explanations, however, are noticeably different.

Student 3 correctly labeled his ratio table to show a comparison of miles to minutes. He generates equivalent ratios to serve as the basis of comparison, though a miscalculation is evident. Student 3 accurately identifies Jessica as the fastest runner, but doesn't explain why.

Like Student 3, Student 4 applies the same strategy and accurately labels the quantities in the ratio table. He correctly determines that Jessica runs 15 miles in 84 minutes and is able to identify Jessica as the fastest runner. Student 4 is able to describe that the fastest runner would have "less time" than the other runner. Student 4's explanation is based on multiplication rather than accurately describing generating equivalent ratios.

*What would we want to ask these students? What might we do next?*

### Student 3

Is Student 3's miscalculation his undoing? Or does it uncover misunderstanding about the ideas of time and distance? We can't be sure how he knows that Jessica is the faster runner as his explanation doesn't share how he knows. We can infer that his thinking is similar to ours, but that can be dangerous. We must be careful of inferring what students mean based on our own understanding. When we do this, we may create understanding that isn't actually in place. It is quite possible, if not likely, that he has a misunderstanding that Jessica is the faster runner because 104 is greater than 90. After all, if he was accurately comparing the two, Jessica would be the faster runner because she ran 15 miles in fewer minutes.

### Student 4

Like Student 3, Student 4 finds a common distance to compare times. His explanation does fairly well to explain how he knows that Jessica's time is faster because it is less than Sara's time. We should continue to help him to refine his justifications so that they are complete. We should also look to see if he is flexible in his approach to solving problems with unit rates. In this problem, multiplying to find a common distance was efficient for him. In fact, he may rely on this approach more often than not. Even so, we also want to be sure he has full understanding and that he is able to divide or compare rates in inverse ways (e.g., miles per minute).

## Student Work 3

Sara ran 3 miles in 18 minutes. Jessica ran 5 miles in 28 minutes. Who is the fastest runner? Explain your reasoning.

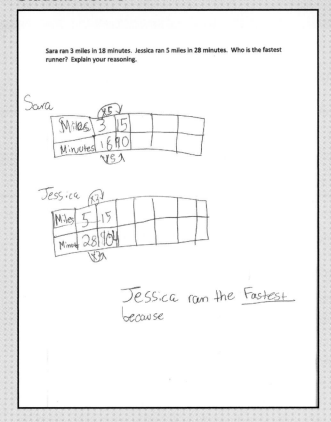

Sara

| Miles | 3 | 15 | | | |
|---|---|---|---|---|---|
| Minutes | 18 | 90 | | | |

Jessica

| Miles | 5 | 15 | | | |
|---|---|---|---|---|---|
| Minutes | 28 | 84 | | | |

Jessica ran the Fastest because

## Student Work 4

Sara ran 3 miles in 18 minutes. Jessica ran 5 miles in 28 minutes. Who is the fastest runner? Explain your reasoning.

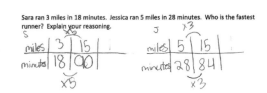

S

| miles | 3 | 15 | |
|---|---|---|---|
| minutes | 18 | 90 | |

×5

J

| miles | 5 | 15 | |
|---|---|---|---|
| minutes | 28 | 84 | |

×3

Jessica is the faster runner because 5×3=15 & 28×3=84 which is less time than what Sarah's time is.

### OTHER TASKS

- What will count as evidence of understanding?
- What misconceptions might you find?
- What will you do or how will you respond?

 Visit this book's companion website at **resources.corwin.com/ minethegap/6-8** for complete, downloadable versions of all tasks.

**TASK 19B:** Circle ALL of the rates below that can be classified as unit rates.

| | |
|---|---|
| $2.00 for 4 oranges | $3.50 for 1 gallon of juice |
| 55 miles per hour | $42 for 7 tickets |
| 3 hours to cook 12 pounds of turkey | $50 per guest |
| 24 players per team | 6 cups of flour for 3 loaves of bread |
| 5 gigabytes of data for each student | 20 notebooks for 10 students |

**For one nonexample, explain how to generate the unit rate from it.**

This task requires students to distinguish between rates and unit rates. A rate is a comparison of two different types of quantities. A unit rate is based on a singular unit of measure for one of the categories. Students may incorrectly believe that a unit rate is only based on a measure of time, such as miles per hour. Other students may only have exposure to unit rates in terms of cost per item, as these are the most common examples in textbooks. Thus, students may overlook examples such as $50 per guest. Students with these misunderstandings may identify examples such as "5 gigabytes of data for each student" incorrectly because the context is contrary to their experiences. To counter this, it is important that we reinforce the meaning of unit rates with a wide variety of contexts and situations.

**TASK 19C:** Describe a scenario that can be represented by a unit rate of $\frac{3}{4}$. Use rate language to explain what the unit rate means in this scenario.

This task provides students with an opportunity to develop a scenario to represent a unit rate of $\frac{3}{4}$. Students may be challenged with a unit rate represented as a fraction, rather than a decimal or whole number. A common type of incorrect response may be a comparison of 3:4, such as $3 for 4 apples. While this is an example of a rate, it does not accurately reflect a unit rate. To accurately represent the unit rate, students need to compare $\frac{3}{4}$ to 1, such as $\frac{3}{4}$ of a gallon for every dollar. Certain contexts will be more helpful for understanding making use of a fraction for a unit rate. It may be wise to confirm that students fully understand the concept of a unit rate before working with situations other than whole number comparisons.

**TASK 19D:** The candy store is having a sale on fudge: 2 pounds of fudge for $10. The store allows you to buy as many or as few pounds of fudge at this same rate. Jamie reads the sign and says, "The cost is $5 per pound." Her sister Janelle reads the sign and says, "No, you get $\frac{1}{5}$ of a pound for every dollar you pay." Who is correct? Explain your reasoning.

It can be helpful to think of rates in different ways depending on the situation or quantities represented by the rates. For example, dollars per pound situations may be better manipulated if thought of as pound per dollar. This task challenges students' understanding of rates. Students must recognize that each sister decided to generate a rate based on a different unit of comparison. Jamie created a unit rate based on a single pound, whereas Janelle created a unit rate based on a single dollar. Students may not see this distinction and thus determine that one or both is not correct. Students may also mistakenly believe that a unit rate cannot be expressed as a fraction, and thus claim that Janelle is incorrect. Students with full understanding understand the relationship and meaning of unit rates and can apply them to either term in the ratio.

NOTES

# BIG IDEA 20
## Using Ratios to Solve Problems

### TASK 20A

> Kara is buying party invitations. The party store sells a $10 box that contains 24 cards or a $25 box that contains 60 cards.
>
> Which is the better deal? Why?
>
> Create another card box that would be equivalent to the better deal.

**MINING TIP**

The task also takes aim at the idea that students assume one ratio must be greater (better) than another in comparison situations.

**MINING HAZARD**

Tape diagrams or double number line diagrams are valuable tools for comparing quantities. Errors may arise as students are determining how to represent the whole. Students may also partition incorrectly, leading to incorrect solutions or comparisons.

### About the Task

Reasoning with ratios is a valuable skill to compare offers and to solve a wide range of problems. This task offers a real-world context for comparing two boxes of cards. The task allows students to use any strategy to determine which box of cards is the better deal and requires students to justify their reasoning. *For this task, students must determine that both options offer the same value per card.* Then, students must apply their understanding of equivalent ratios in order to design another offer that is equivalent to the better deal.

### Anticipating Student Responses

Students may approach this task using a variety of strategies. Students may try to compare the unit rates for the boxes, by either determining the cost per card or the number of cards that may be purchased for $1. *Some students may choose to use tape diagrams or double number line diagrams to compare.* Some students may use a ratio table to find equivalent box offers. For example,

**PAUSE AND REFLECT**

- How does this task compare to tasks I've used?

- What might my students do in this task?

Visit this book's companion website at **resources.corwin.com/minethegap/6-8** for complete, downloadable versions of all tasks.

the student may use the relationship $10 box for 24 cards to determine what the cost would be for 60 cards, then compare that value to $25. Another common strategy is to find the greatest common factor (GCF) of one of the units then multiply by that factor. Students may find that the GCF of 10 and 25 is 50. They will find the number of cards based on this common factor.

NOTES

### Student 1

Student 1 attempts to generate equivalent ratios as a strategy for comparison. She increases the money by the same amount each time, increasing by $10 in the first table and by $25 in the second table. She adjusts the number of cards accordingly. Student 1 doesn't expand the first table to examine the number of cards that may be purchased for $50. Student 1 then determines that the second deal is better, but provides no explanation for this conclusion.

### Student 2

Student 2 attempts to create a ratio table to compare the two offers. She attempts to find the cost to purchase 1 card in each of the tables. Student 2 chooses to multiply rather than divide to generate the costs. She sees a relationship between 2.4, 10, and 24 and notes it. However, she doesn't reason that $2.40 \times 24$ would have a product greater than 48. She then uses faulty values for the basis of comparison. Student 2 then creates another offer, but it is unclear how she generates the cost and number of cards purchased.

***What would we want to ask these students? What might we do next?***

### Student 1

We can build on Student 1's understanding. She shows that she can generate equivalent ratios by increasing both values by the same factor. She is still developing a valid strategy for comparing two ratios such as those in this task. We should first have Student 1 articulate why she selected the second offer. This understanding can help us determine other steps. Regardless, we'll need to provide new opportunities to compare ratios in varied contexts. Each time, it will be essential to highlight that one of the values (cards or money) must be equivalent so that we can draw a conclusion.

### Student 2

Student 2 has a strategy for comparison but has made computational errors. We might address the error by asking the student how she knew to multiply in each table. We can ask her to describe how she multiplied the money and the cards. We need to draw attention to her $2.40 \times 24$ and $15 \times 60 = 25$. We could simply ask what is $2 \times 24$ or $15 \times 2$? This should help her recognize the error. Hopefully, this better establishes reasoning rather than reliance on presumed accurate calculations. Knowing just this should be evidence that the task needs to be reconsidered. Student 2 is a good example of the importance of developing reasoning and mental mathematics skills as well as the value in comparing solutions to other works.

## TASK 20A: Kara is buying party invitations. The party store sells a $10 box that contains 24 cards or a $25 box that contains 60 cards. Which is the better deal? Why? Create another card box that would be equivalent to the better deal.

### Student Work 1

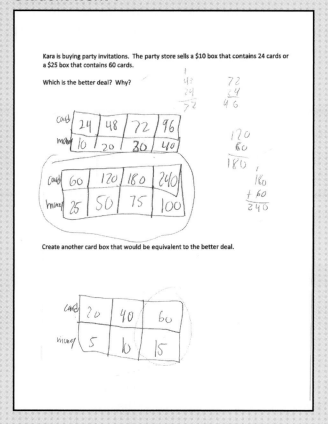

### Student Work 2

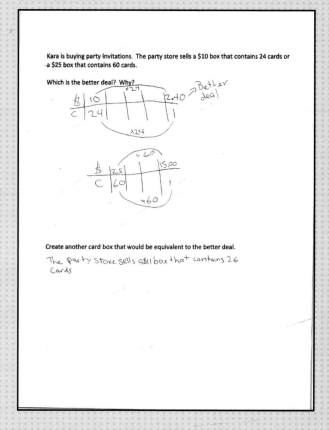

### Student 3

Student 3 uses a viable strategy to generate a basis for comparison between offers. She generates a unit rate for cost based on the purchase of one card by dividing. She determines that each card costs about $0.42 in both offers and notes that "they both are the same amount." Student 3's answer doesn't satisfy the prompt. She establishes correctly that the two are equal. She shows that she can find a unit rate for the new box but doesn't consider its relationship to the found unit rate.

### Student 4

Student 4 appears to have selected the smaller box of cards first, but then decides to use an alternate strategy. In the revised strategy, she creates a table in which the cost increases by $10 as the number of cards increases by 24. She realizes that in each set you can buy 120 cards for $50. Student 4 generates a scenario in which the cost per card is not equivalent to the other offers. Interestingly, she selects a dollar amount that is already represented in the table for 60 cards.

*What would we want to ask these students? What might we do next?*

### Student 3

Student 3 shows that she understands how to use unit rates to compare offers. Her error in the extension may be attributed to different reasons. Our next steps depend on the cause of the error. She may have misread or misunderstood the prompt. She may be unable to evaluate the equivalency of more than two ratios. If so, our subsequent work should focus on just that. Or, she may be unable to generate equivalency through unit rates. This challenge can be indicative of a student who has learned to *do* math rather than one who has learned to *know* or understand math. If this is the case, we should place a renewed emphasis on the concept and representations of it.

### Student 4

Like Student 3, Student 4 is able to use an appropriate strategy for comparing two offers. It is not clear whether Student 4 would have the same success if the common cost was not as easy to obtain. We should provide these opportunities and reinforce the strategy of finding a unit rate. Related to this task, we might ask her to find the number of cards in boxes selling for $21. We also want to investigate her generated response. We should connect the $25 price of the second box in the prompt, asking how the conditions are the same or different. It is possible that her error is not with the mathematics. Simply, $25 for 75 cards *is* a "better deal," meaning she overlooked the "equivalent to" portion of the prompt.

## Student Work 3

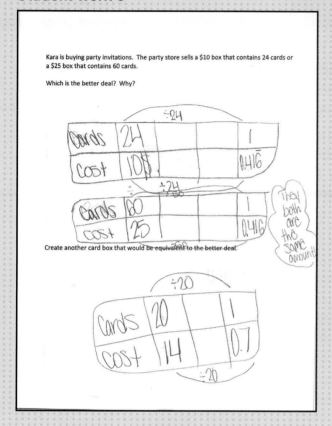

Kara is buying party invitations. The party store sells a $10 box that contains 24 cards or a $25 box that contains 60 cards.

Which is the better deal? Why?

Create another card box that would be equivalent to the better deal.

## Student Work 4

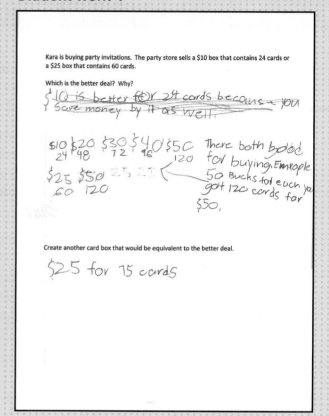

Kara is buying party invitations. The party store sells a $10 box that contains 24 cards or a $25 box that contains 60 cards.

Which is the better deal? Why?

$10 is better for 24 cards becouse you save money by it as well.

$10 $20 $30 $40 $50    There both good
24  48  72  96  120    for buying. Example
$25 $50  25 2.5        50 Bucks for each ya
60  120                got 120 cards for
                       $50.

Create another card box that would be equivalent to the better deal.

$25 for 75 cards

## OTHER TASKS

- What will count as evidence of understanding?

- What misconceptions might you find?

- What will you do or how will you respond?

 Visit this book's companion website at **resources.corwin.com/ minethegap/6-8** for complete, downloadable versions of all tasks.

**TASK 20B:** Jayden scored a 42 out of 60 on his English test. Imani has a different English teacher and scored a 20 out of 25 on her test. Who did better on their test? Why?

Red Hill Middle School's mural of the school mascot is 12 feet by 16 feet. Mariah drew a smaller version of the mural for a t-shirt design. Her mascot mural is 9 inches by 12 inches. Is Mariah's mural proportional to the school mural? Explain your reasoning.

This task requires students to compare ratios in two different real-world contexts. In both prompts, students may use a variety of strategies to solve, including ratio tables, unit fractions, tape diagrams, and finding other equivalent ratios. Some students may argue that Jayden did better because he scored more points. These students show that they aren't considering the relationship between points scored and points possible. We may also find that students correctly identify Imani but for the wrong reasons. These students may hold misconceptions about ratios as additive relationships and respond based on which difference between quantities is greatest. The second prompt may provide greater insight into our students' reasoning about mathematics and situations. The two murals are in different units of measurement. Our students may be compelled to convert to the same unit before finding if the ratios are equivalent. More advanced students will argue that the rates can be compared because the units are the same in each ratio.

**TASK 20C:** The table shows the ratio of goals to shots on goal for two hockey teams in a five-game series. Circle the team that had the greater ratio of goals to shots for each game. Choose one of the games you compared above. Tell how you found the team with the greater ratio of goals to shots.

|  | Sharks (goals to shots) | Bears (goals to shots) |
|---|---|---|
| Game 1 | 1:13 | 2:17 |
| Game 2 | 3:31 | 1:10 |
| Game 3 | 2:6 | 4:15 |
| Game 4 | 3:29 | 2:9 |
| Game 5 | 3:18 | 1:7 |

This task affords students opportunities to apply understanding of equivalent fractions and multiples to compare ratios. Students may use a variety of strategies to solve. This is an ideal problem for determining whether students are beginning to use efficient strategies to solve problems. Since some of these ratios involve 1 goal, proficient students may recognize that the most efficient strategy is to double or triple the goals to compare. For example, 1:13 is equivalent to 2:26, which is a poorer goal-to-shot ratio compared to 2:17. A similar strategy might be applied to game 3. With it, students could find that 4:12 is equivalent to 2:6. They can then compare 4:12 to 4:15, noting that 4:12 or 2:6 is greater than 4:15.

## TASK 20D: Josh and Samantha are planning to have a new home built. First, they need to select one of the lots of land to purchase. Here are the available lots:

**Lot 9: 250 feet by 258 feet**

**Lot 13: 220 feet by 275 feet**

**Lot 15: 240 feet by 260 feet**

**Samantha would like the lot to be as square as possible. Which lot should they purchase? Explain your reasoning.**

Task 20D provides another opportunity to examine strategies used to solve problems. *This task also connects to key understandings of figures and area.* Students may be more likely to apply the incorrect reasoning of finding the difference between plot dimensions or area rather than comparing the ratios between the lengths and widths. Incidence of this error increases when this task is given outside of the study of ratios. Thus, it is important to continue to monitor student understanding of ratio reasoning as we progress into other topics of study. At first glance, the task may seem to require considerable calculations. However, if understanding of a square's dimensions as being a ratio of 1:1 is in place, the task becomes less complicated. With this understanding, we can see that Lot 9's dimensions have the smallest difference. As we know, the closer these dimensions are to one another, the closer the rectangle is to being a square.

**CONTENT CONNECTION**

While this task is assessing understanding of comparing ratios, it is also connected to geometry concepts of area and properties of a square. We should be sure that our students have an understanding of the vocabulary and key concepts prior to working with this task.

NOTES

# BIG IDEA 21
## Reasoning With Percents

### TASK 21A

Compare the rows. Circle the greater amount.

| | |
|---|---|
| 60% of 99 | 50% of 199 |
| 75% of 80 | 80% of 90 |
| 25% of 950 | 90% of 200 |
| 30% of 15 | 150% of 12 |

Choose one of the rows. Tell how you determined the greater amount.

## About the Task

We can calculate to find exact values. Often, reasoning about numbers and situations is much more efficient. Ample opportunities to reason about relationships or calculations can improve our students' accuracy as they develop ideas about reasonableness of their answers. This is especially helpful when their calculations are incorrect. This task provides students an opportunity to use a variety of strategies to compare percents of numbers. The prompts include landmark percents as well as percents that exceed 100%. This task provides an opportunity to connect with previous understandings of fractions and decimals that can be applied to similar reasoning with percents.

## Anticipating Student Responses

We will have students who compute to justify their comparisons. Other students will apply different types of strategies to make comparisons. *Some students*

### PAUSE AND REFLECT

- How does this task compare to tasks I've used?

- What might my students do in this task?

Visit this book's companion website at
**resources.corwin.com/minethegap/6-8**
for complete, downloadable versions of all tasks.

*who have learned how to set up a proportion may solve using a proportion.* We may find that students apply understanding in different ways. For example, they might reason that 80% of 90 is greater than 75% of 80 because both (80% and 90) are greater. Others may reason that 30% of 15 yields something less than 7.5 because it is 50% of 15 while 150% of 12 will be more than 12. To compare the first line, students might note that 50% of 199 is about 100 and so that must be greater than 60% of 99 because 99 is less than 100. Students may mistakenly assume that the larger quantity will always yield a larger quantity. Similarly students may overgeneralize and assume that the larger percent will also yield the larger quantity, regardless of the original value.

**MINING HAZARD**

While using a proportion is a viable strategy, this strategy does not necessarily guarantee that students have a conceptual understanding of reasoning with percents. Student explanations can help us understand how to proceed.

NOTES

### Student 1

Student 1 understands that percents represent portions of 100. He overgeneralizes this understanding to assume that the larger the percentage, the larger the quantity will be. This reasoning could only be applied if different percents were applied to the same original quantity. He doesn't recognize that 199 is approximately twice as large as 99, so taking one half of that quantity will result in a number close to 100.

**MINING HAZARD**

**Correct answers can lead us to believe that a student's strategy is fine. It is critical to assess the reasoning to ensure that students have a conceptual understanding of why the strategy works.**

### Student 2

Student 2 has correctly identified which amount is greater. He has applied the procedure of setting up a proportion in order to compare the quantities. *While this procedure effectively yields the solution, the explanation does not reveal whether the student understands how the percent is affecting the original quantity.* He hasn't communicated understanding of how he can reason about a percent of quantities.

*What would we want to ask these students? What might we do next?*

### Student 1

When working with this student, we can build on his understanding of what a percent is. The next step is to bring attention to the need to compare the original quantities. His challenge may be rooted in a misunderstanding about the changing size of a whole with fractions. Since this student referenced using fractions, have him estimate what half of each of these quantities is. We should apply his use of a number line using the quantities 99 and 199 or near quantities of 100 and 200 as endpoints.

### Student 2

Student 2 has a procedural strategy that works well for him. Yet, procedure is not always efficient. We must begin to shift his approach to reasoning and relying on number sense. We might ask him to use a different strategy to confirm his solution. One way to do that is to ask, "How could you check your solution by examining the quantities and percents?" If number sense is still developing, begin with the same original quantity, such as 60% of 99 compared to 50% of 99 before comparing different quantities. We can work with these ideas as a specific lesson or as a brief routine we might use daily.

**TASK 21A:** Compare the rows. Circle the greater amount.

Choose one of the rows. Tell how you determined the greater amount.

## Student Work 1

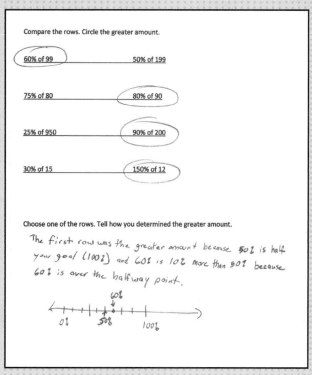

Compare the rows. Circle the greater amount.

(60% of 99)          50% of 199

75% of 80          (80% of 90)

25% of 950          (90% of 200)

30% of 15          (150% of 12)

Choose one of the rows. Tell how you determined the greater amount.

The first row was the greater amount because 50% is half your goal (100%) and 60% is 10% more than 50% because 60% is over the halfway point.

## Student Work 2

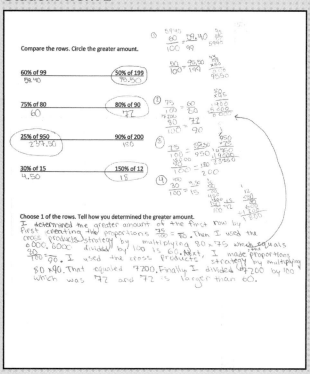

Compare the rows. Circle the greater amount.

60% of 99          (50% of 199)
59.40                 99.50

75% of 80          (80% of 90)
60                      72

(25% of 950)          90% of 200
237.50                 180

30% of 15          (150% of 12)
4.50                    18

Choose 1 of the rows. Tell how you determined the greater amount.
I determined the greater amount of the first row by first creating the proportions 75/100 = 60/80. Then I used the cross products strategy by multiplying 80 × 75 which equals 6000. 6000 divided by 100 is 60. Next, I made proportions 80/100 = 72/90. I used the cross products strategy by multiplying 80 × 90. That equaled 7200. Finally I divided 7200 by 100 which was 72 and 72 is larger than 60.

### Student 3

Student 3 has an appropriate estimation strategy to determine the greater amount. He sets up proportions to find the exact quantity. We can see that he, like Student 2, uses calculations to find these values in order to compare them. He justifies his approach with row 2 likely due to the values in the row. Of the four rows, row 2 has the cleanest computations as it features percents in multiples of 5 and whole numbers in multiples of 10.

### Student 4

**MINING TIP**

Student 4 is a good example of why it is necessary for students to explain their reasoning. His misconception may have lingered or even strengthened had it not been uncovered here.

Though the comparisons are correct, Student 4's response shows that he considers more reasonable numbers when working with these situations. *It also reveals significant flaws.* The student appears to understand that percents relate to 100, but does not show any understanding of how percents of a quantity work. He also is relying heavily on additive relationships as he subtracts 1 from 40 (which he intended to be 90) and he subtracts 1 from 50% of 200 because it is 1 more than 199. His work for row 2, though accurate, conflicts with his approach in row 1.

**USING EVIDENCE**

*What would we want to ask these students? What might we do next?*

### Student 3

Thinking about percents proportionally is a good strategy for finding a percent of a value. This reasoning can be applied to estimated or benchmark values. Doing so improves efficiency. We can help Student 3 make this connection by discussing why he justified row 2. It is likely that he will talk about the relative ease of those calculations. We can then shift the conversation to highlight how we might think of the unfriendly values in more friendly ways. Essentially, Student 3 has the building blocks to be a more efficient reasoner. Our work is to help him connect those ideas and provide opportunities to reinforce them.

### Student 4

We should renew efforts with Student 4 with work focused on developing a conceptual understanding of percents, decimals, and fractions. We need to revisit percent as a quantity out of 100, rather than "away from 100." Visual representations and connections to fractions will help with building conceptual understanding. As this understanding takes shape, we can shift to develop an understanding of what it means to take a percent of a quantity. Tape diagrams or double number line diagrams can be used to represent each problem to show what that problem means. As we work with these representations, he can begin to make comparisons between percents of quantities.

**TASK 21A:** Compare the rows. Circle the greater amount.

Choose one of the rows. Tell how you determined the greater amount.

### Student Work 3

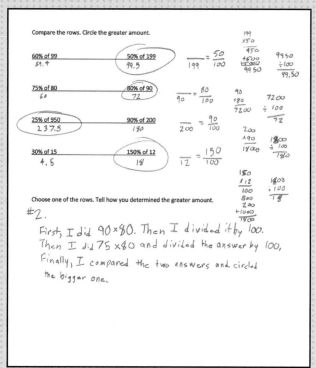

Compare the rows. Circle the greater amount.

60% of 99      (50% of 199)
59.4      99.5

75% of 80      (80% of 90)
60      72

(25% of 950)      90% of 200
237.5      180

30% of 15      (150% of 12)
4.5      18

Choose one of the rows. Tell how you determined the greater amount.

#2.

First, I did 90×80. Then I divided it by 100.
Then I did 75×80 and divided the answer by 100,
Finally, I compared the two answers and circled
the bigger one.

### Student Work 4

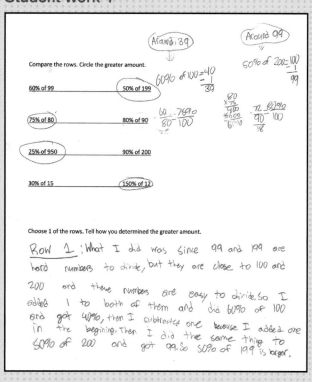

Compare the rows. Circle the greater amount.

60% of 99      (50% of 199)

(75% of 80)      80% of 90

(25% of 950)      90% of 200

30% of 15      (150% of 12)

Choose 1 of the rows. Tell how you determined the greater amount.

Row 1: What I did was since 99 and 199 are
hard numbers to divide, but they are close to 100 and
200 and these numbers are easy to divide. So I
added 1 to both of them and did 60% of 100
and got 40%, then I subtracted one because I added one
in the begining. Then I did the same thing to
50% of 200 and got 99. So 50% of 199 is larger.

Visit this book's companion website at resources.corwin.com/ minethegap/6-8 for complete, downloadable versions of all tasks.

**TASK 21B:** Determine if each is greater or less than 25%.

| | | |
|---|---|---|
| 25 ÷ 99 | greater than 25% | less than 25% |
| 130 ÷ 1,320 | greater than 25% | less than 25% |
| 91 ÷ 899 | greater than 25% | less than 25% |
| 29 ÷ 1,608 | greater than 25% | less than 25% |
| 33 ÷ 142 | greater than 25% | less than 25% |

**Choose one of the problems. Use models, numbers, or words to explain how you knew the value was greater or less than 25%.**

Task 21B requires students to make connections between fractions and percents. Students are provided the landmark decimal of 25% and should estimate which of the quantities produced from the division problem will result in a value that is greater than or less than 25%. Proficient students will look for ways to quickly estimate the fraction decimal without performing the division. For example, students should recognize that 25 ÷ 100 is 25% so 25 ÷ 99 must be greater than 25%, relating to a common numerator strategy for comparing fractions. In fact, the other situations can be estimated as well. 130 ÷ 1320 is close to 130 ÷ 1300, which is a friendlier computation. The same could be applied to 91 ÷ 899, which is about 90 ÷ 900.

**TASK 21C:** Tell if the result is greater or less than 50.

| | | |
|---|---|---|
| 30% of 200 | greater than 50 | less than 50 |
| 10% of 625 | greater than 50 | less than 50 |
| 99% of 40 | greater than 50 | less than 50 |
| 75% of 50 | greater than 50 | less than 50 |
| 24% of 250 | greater than 50 | less than 50 |

**MODIFYING THE TASK**

We can easily modify this task by changing the values in it. We should be sure to make use of benchmark numbers and percents. We can have students create examples for us to pose to the class for discussion.

**Choose two of the rows above and explain how you found your comparison.**

Task 21C is another take on reasoning about percents and values. *This task requires students to estimate the percent of a number and then compare that quantity to 50.* We might use this task before task 21A for students who need additional support with understanding what a percent of a quantity means. This task can help students build number sense by reasoning about the quantities. Some students may choose to represent with landmark numbers or use tape diagrams or other representations to determine solutions. Proficient students may begin by estimating what fraction 50 out of the quantity would be and then comparing to the percent. For example, 50 is 25% of 200, so 30% of 200 would produce a value that is greater than 50.

**TASK 21D:** Kristin purchased two identical pairs of boots for her twin daughters. She paid a total of $84 after receiving a 30% discount. What was the original price of one pair of boots?

**Leila earned a score of 70% on a quiz. She answered 28 questions correctly. If each question was worth 1 point, how many questions were on the quiz?**

**During the Labor Day sale, the outlet store offers a 40% discount on all items in the store. If the original price of a purse is $240, how much will it cost after the discount?**

Each of the three problems in this task requires students to apply what they know about percents of quantities to solve real-world problems. The second problem is a percent problem that requires students to find the original amount. Students may misread this problem and find 70% of 28. The first and third questions deal with a percent discount. For these problems, students may simply take the percent of the quantity and not realize that they found the amount discounted, rather than the final price or original price. Proficient students may realize that they need to subtract the percent from 100% to find the percent that they will need to pay for the item. Other students may find the discounted amount and then subtract from the original price. As we know, these types of problems can be quite challenging for students. *It is wise to begin with friendly numbers and percents to first determine if students understand how to solve the problems before working with more complicated values.*

**MINING HAZARD**

The values we use in problems may create barriers for students. The complexity of computations may undermine our attempts to determine student understanding.

NOTES

# BIG IDEA 22
## Unit Rate as Slope

### TASK 22A

> For 5 hours of work, Macey earned $45.
>
> Assuming she earned a constant rate per hour, create a table, graph, and/or equation to display her earnings for 1 hour, 5 hours, and 10 hours of work.
>
> Explain the relationship between the number of hours and the amount of money earned.

### About the Task

Understanding unit rate as slope is a foundational understanding for the study of algebra and functions. This key understanding builds from an understanding of the standards connected to ratios and proportional relationships. For this task, students will begin to connect their understanding of unit rates to different representations of this relationship, including a table of values, a graph, and an equation. Students will begin to see that the ratio between any two points is constant. This relationship can also be represented as the direct variation equation $y = kx$, where $k$ is the unit rate.

### Anticipating Student Responses

In order to represent this unit rate with a table, graph, or equation, students must first calculate the unit rate of $9 per hour. Students should be able to translate this unit rate to find the earnings for 1 hour and 10 hours, either by

### PAUSE AND REFLECT

- How does this task compare to tasks I've used?
- What might my students do in this task?

Visit this book's companion website at **resources.corwin.com/minethegap/6-8** for complete, downloadable versions of all tasks.

using the unit rate or by creating the table of values as a ratio table. *Translating this relationship to a graph or equation requires students to identify variables to represent the hours worked and total earnings.* We may have students who find the amount earned for other hours to find the specific amounts in the prompt. These and other students may have difficulty writing an equation to represent the situation or may assign the variables to the wrong axes.

**CONTENT CONNECTION**

This task begins to connect the ratios and proportional relationships understanding to expressions and equations standards. In particular, students need to have an understanding of how to plot points on the coordinate plane, write equations to represent situations, and distinguish between independent and dependent variables.

NOTES

## Student 1

Student 1 generates a table of values that shows the earnings increase by $9 as the hours increase by 1. She has labeled earnings as the independent variable and hours as the dependent variable. She knows a formula for the constant of proportionality and shows the equation $y = 9x$, but the student is not able to recognize that the equation does not match how she has defined the variables. Student 1's graph also plots the earnings on the $x$-axis and the time on the $y$-axis. Student 1 doesn't explain the relationship between the number of hours and the amount of money earned.

## Student 2

Student 2 accurately determines the earnings for 1, 5, and 10 hours of work. Like Student 1, Student 2 represents the earnings on the $x$-axis and the hours worked on the $y$-axis. This student also uses the labels "per hour" and "per dollar," indicating her misunderstanding of what "per" means. Student 2 accurately states, "This graph presents proportionality in this instance." Yet, it is not clear whether the student understands *why* this represents a proportional relationship or what the specific relationship between the number of hours and total earnings is.

## USING EVIDENCE

### *What would we want to ask these students? What might we do next?*

## Student 1

Student 1 shows that there is a constant increase of $9 each hour. However, she doesn't explicitly connect her representations. We can ask her to explain the relationship between the number of hours and money earned. Then, we can ask her how each of the representations models this relationship. We should be sure to focus on the equation and how she has defined the variables. We might have her substitute one of the earnings for $x$ to find the output so she can discover her error and reconcile it.

## Student 2

Student 2 has a representation and shows some understanding of the relationship in this task. We need to be sure that she understands independent and dependent variables and that "per" has a clear meaning. Unlike Student 1, she features only one representation. This is not necessarily problematic but could indicate misunderstanding or discomfort with other examples. She may think of a graph as *her* only option. Student 2 will benefit from exposure to other students' representations and approaches. We might think to strategically place her in a group for a similar task to do this. As with Student 1, we will need to be sure that she understands how the representations prove the proportional relationship. We also need to ensure that Student 2 understands how the two variables relate to one another.

## TASK 22A: For 5 hours of work, Macey earned $45. Assuming she earned a constant rate per hour, create a table, graph, and/or equation to display her earnings for 1 hour, 5 hours, and 10 hours of work. Explain the relationship between the number of hours and the amount of money earned.

## Student Work 1

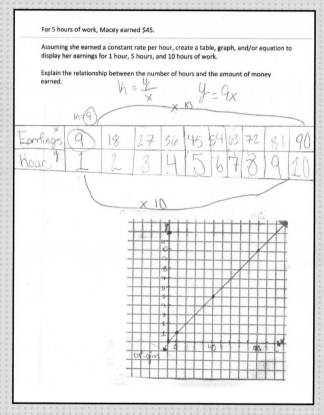

For 5 hours of work, Macey earned $45.

Assuming she earned a constant rate per hour, create a table, graph, and/or equation to display her earnings for 1 hour, 5 hours, and 10 hours of work.

Explain the relationship between the number of hours and the amount of money earned.

$h = \frac{y}{x}$    $y = 9x$

$h = 9$

| Earnings | 9 | 18 | 27 | 36 | 45 | 54 | 63 | 72 | 81 | 90 |
|----------|---|----|----|----|----|----|----|----|----|----|
| Hours | 1 | 2 | 3 | 4 | 5 | 6 | 7 | 8 | 9 | 10 |

× 10

Origin

## Student Work 2

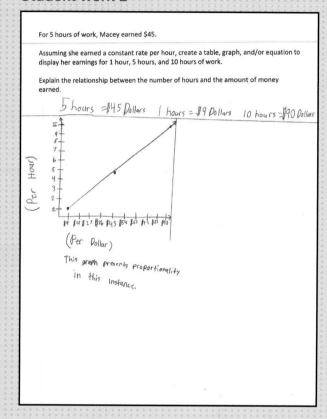

For 5 hours of work, Macey earned $45.

Assuming she earned a constant rate per hour, create a table, graph, and/or equation to display her earnings for 1 hour, 5 hours, and 10 hours of work.

Explain the relationship between the number of hours and the amount of money earned.

5 hours = $45 Dollars    1 hours = $9 Dollars    10 hours = $90 Dollars

(Per Hour)

(Per Dollar)

This graph presents proportionality in this instance.

## Student 3

Student 3 generates three tables of values. In her "*xy*" table, she shows that the hours represent the independent variable and the earnings represent the dependent variable. Student 3 also shows that when you divide the output by the input value, the result is a constant of 9. Student 3 concludes that the constant of proportionality is 9. She doesn't explicitly connect the number of hours and the amount of money earned.

## Student 4

Like Student 3, Student 4 accurately generates a table of values and defines the number of hours as the independent variable. Student 4 shows that to generate the earnings, she must multiply the number of hours by 9. She produces an accurate graph in which the hours are plotted on the *x*-axis and the earnings are plotted on the *y*-axis. She describes that "no matter how much money or hours worked it will always be $9 an hour."

*What would we want to ask these students? What might we do next?*

## Student 3

Student 3 has an understanding of how to generate a table of values using equivalent ratios. It is not clear whether the student can then translate this relationship in the context of the scenario. We can ask Student 3 what the relationship between the number of hours and the amount of money earned is. Assuming her response is accurate, we can ask how this relationship connects to the constant of proportionality. As we know, graphical relationships are useful in many situations. Our work with Student 3 is to help her connect and reinforce her varied table relationships to graphs of these relationships.

## Student 4

Student 4 represents proportional relationships as a table of values and as a graph. She understands the constant of proportionality in this context is the $9 per hour. There are ways to advance Student 4's understanding. We can ask her to determine an equation to represent this relationship and ask her to describe how each representation, including the equation, represents a proportional relationship. We might also extend Student 4 by providing different situations with similar contexts to compare. Another option may be to present situations for her to consider if a proportional relationship exists at all.

## Student Work 3

For 5 hours of work, Macey earned $45.

Assuming she earned a constant rate per hour, create a table, graph, and/or equation to display her earnings for 1 hour, 5 hours, and 10 hours of work.

Explain the relationship between the number of hours and the amount of money earned.

Macey's earnings

| 1 hr | 5 hr | 10 hr |
|------|------|-------|
| 9 | 45 | 90 |

| money | 9 | 45 | 90 |
|-------|---|----|----|
| hour | 1 | 5 | 10 |

| X | Y | |
|---|---|---|
| 1 | 9 | $\frac{9}{1}$ |
| 5 | 45 | $\frac{45}{5}$ |
| 10 | 90 | $\frac{90}{10}$ |

Yes their is a constant of proportionality its 9.

## Student Work 4

For 5 hours of work, Macey earned $45.

Assuming she earned a constant rate per hour, create a table, graph, and/or equation to display her earnings for 1 hour, 5 hours, and 10 hours of work.

Explain the relationship between the number of hours and the amount of money earned.

1) $1 \times \$9 = \$9^1$
2) $5 \times \$9 = \$45 = \$9^5$
3) $10 \times \$9 = \$90 = \$9^{10}$

Etc

| Hour(x) | Money(y) |
|---------|----------|
| 1 | 9 |
| 5 | 45 |
| 10 | 90 |

Macey earns $9 an hour. in 5 hours she has gotten $45. Know matter how much money or hours worked it will alway be $9 an hour.

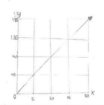

## OTHER TASKS

- What will count as evidence of understanding?
- What misconceptions might you find?
- What will you do or how will you respond?

 Visit this book's companion website at **resources.corwin.com/minethegap/6-8** for complete, downloadable versions of all tasks.

**TASK 22B:** Devon runs 8 miles in 48 minutes.

**If he runs at a constant rate, how long does it take him to run a mile?**

**How long would it take him to run a marathon (26.2 miles)?**

**MODIFYING THE TASK**

To extend this problem, have students represent this problem using a graph or equation.

This task requires students to apply understanding of unit rates and proportional reasoning to find equivalent ratios in a real-world situation. Many students will first find the unit rate to find that it will take 6 minutes to run 1 mile. Most students will then multiply the 26.2 miles by 6 to find how long it will take to run a marathon. Student errors may arise if students find a unit rate for 1 minute rather than 1 mile. Miscalculations are also likely to occur. Even so, we should be able to see a strategy for finding a unit rate and applying it to the problem. *Students without these strategies likely need more work developing understanding of unit rates.*

**TASK 22C:** The school SGA is planning a field trip to the local amusement park. A value ride package for four people costs $54. The SGA wants to determine how much to charge students for the field trip.

**How much should each student pay?**

**How much does the SGA need to collect if there are 120 students attending the field trip?**

This task provides another opportunity to find and apply understanding of unit rates in order to solve real-world problems. In this task, students need to determine the cost per person and then use that rate to determine the amount of money needed for 120 students to attend the field trip. Most students will calculate the unit rate first, then multiply the unit rate by 120 to find the total needed. Some students may choose to use a ratio table to find the amount to collect for 120 students to attend the trip. Students with a deep understanding of unit rate as slope may write and use the equation $C = 13.5p$, where $p$ represents the number of people attending and $C$ represents the total cost, in order to find the total cost for 120 people. As with other tasks, this task is a good opportunity for students to work with partners or in groups before bringing the class together to discuss strategies and solutions. Doing this exposes students to other approaches and can help reinforce understanding of unit rates as well as more efficient approaches.

**TASK 22D:** Find the unit rate for each ticket deal in the graph.

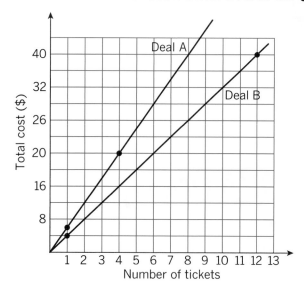

## Which is the better deal? Explain your reasoning.

This task requires students to apply their understanding of unit rates as slope in order to compare two linear functions. Students will need to determine the unit rate, or slope, of each line and then determine that, in this case, the smaller unit rate will represent the better deal. Common errors students may make include miscalculating the unit rates by reversing the fraction (tickets per dollar rather than cost per ticket) or by neglecting to account for the different scales on the *x*- and *y*-axes. Students may also incorrectly state that Deal A is the better deal simply because the unit rate is larger, rather than considering the context for the task. To address this misconception, ask students to create a table of values for some sample numbers of tickets and use the table to verify their conjecture.

NOTES

## CHAPTER 5

# EXPRESSIONS, EQUATIONS, AND FUNCTIONS

**THIS CHAPTER HIGHLIGHTS HIGH-QUALITY TASKS FOR THE FOLLOWING:**

- Big Idea 23: Writing Expressions

  Expressions are useful models for solving problems. We can write expressions to describe quantities or situations.

- Big Idea 24: Evaluating Expressions

  The value of an expression changes as the value of the variable or variables within that expression change. Evaluating expressions lays the foundation for understanding inputs and outputs associated with functions.

- Big Idea 25: Equivalent Expressions

  Expressions can be equivalent even if they look very different. We can make use of a variety of strategies for determining if two expressions are equivalent.

- Big Idea 26: Writing Equations

  Equations are symbolic models of equivalent values. We can write equations to solve problems.

- Big Idea 27: Solving Equations

  Equations enable us to find values and solve problems. We can use a variety of strategies for solving equations.

- Big Idea 28: Inequalities

  Inequalities compare values, contexts, or situations. Solving inequalities results in a range of possible solutions for specific situations.

- Big Idea 29: Function Tables

  Function tables provide inputs and outputs that may be modeled by a rule or function. We can use function rules to find specific inputs and outputs as well as describe the type of function modeled.

- Big Idea 30: Reasoning About Graphing

  Graphical representations of relationships help us describe situations, solve problems, and draw conclusions.

- Big Idea 31: Comparing Functions

  Comparing functions is grounded in understanding of functional relationships. We can compare functions to make predictions, draw conclusions, and solve problems.

- Big Idea 32: Systems of Equations

  A system of equations is a group of two or more equations with the same set of unknowns. Understanding how they work enables us to use systems to solve real-world problems.

NOTES

# BIG IDEA 23
## Writing Expressions

### TASK 23A

Mr. Kropach asks his class to write an expression for the following scenario:

A parking garage charges $4 each hour, but parking is free for the first hour.

- Tamra wrote the expression $4h$, where $h =$ the number of hours the car is in the garage.

- Pratik wrote the expression $4(h - 1)$, where $h =$ the number of hours the car is in the garage.

- Luis wrote the expression $4h - 4$, where $h =$ the number of hours the car is in the garage.

Who is correct? Why?

### About the Task

Writing expressions is a critical skill that serves as a foundation for algebraic reasoning. Expressions help us generalize the relationships in changing situations. However, writing expressions can be challenging as our students develop reasoning and understanding. This task provides the opportunity for students to analyze whether different expressions represent a given scenario. Students have the opportunity to consider if different expressions could be equivalent. The task also presents a real-world context for an expression requiring the use of parentheses.

### PAUSE AND REFLECT

- How does this task compare to tasks I've used?

- What might my students do in this task?

 Visit this book's companion website at
resources.corwin.com/minethegap/6-8
for complete, downloadable versions of all tasks.

## Anticipating Student Responses

Our students are likely to approach this problem in a few different ways. Students may begin with the scenario and write one expression, then look to see if any of the students in the example wrote the same solution. For example, a student may write $4(h - 1)$ and then select Pratik's response. These students may fail to consider the other choices because they have found their one correct solution. *Some students may neglect the free hour of parking and simply select Tamra's expression because the garage charges $4 per hour.* Other students may test input values or create a table of values for each expression to determine whether the costs would be correct. If the car was parked for 3 hours, the charge would be $8. Then the student would look to determine which expression(s) would yield this output.

**MINING HAZARD**

When writing expressions, it is important that students understand that the output is a result of the entire scenario. By omitting the hour of free parking, students have not fully developed an understanding of the purpose of the expression.

NOTES

### Student 1

Student 1 incorrectly identifies Tamra's expression as the correct expression. He only considers how to address the hourly charge and does not account for the free hour of parking. In his description, Student 1 states that "we do not know how long he stayed so the free parking is useless in this equation." This is evidence that he is not considering that the total charge also includes the discount for the first hour. We also note he misuses the term "equation" in place of "expression."

### Student 2

Student 2 accurately identifies Pratik's expression as a correct solution. He is able to articulate that one hour must be subtracted to represent the free hour of parking. Student 2 doesn't consider that there may be more than one correct expression. This may be evidence of unfinished learning about various ways to represent equivalent expressions.

## USING EVIDENCE

*What would we want to ask these students? What might we do next?*

### Student 1

We first want to be sure that Student 1 understands the difference between an equation and an expression. We can ask him to find the total charge for 1, 3, and 10 hours of parking with and without the first hour discount to highlight his misunderstanding about the free parking. We should help him connect his ideas with an expression. Another approach could be to have him find a solution for a specific amount of time and then use that value to evaluate each expression so that he can compare his solution with the values each expression creates. For example, we might have him find the cost for 3 hours ($8). Then, he could substitute and evaluate each expression to determine which is equal to 8.

### Student 2

Student 2 has accurately explained why Pratik's expression is correct, but shows why it is important to dig into student understanding. It may be that he approaches problems with the intent to find one solution. Or, he may understand how to find equivalent expressions but fail to recognize them in open, problem-based contexts. We also want him to consider the other expressions. We might ask him how he knows that the other expressions can't represent the scenario. We may encourage him to find a solution for any number of hours with his chosen expression before evaluating the other expressions with that same number of hours. It will be important to provide more opportunities for him to connect situations to more than one expression.

**TASK 23A:** Mr. Kropach asks his class to write an expression for the following scenario:

A parking garage charges $4 each hour, but parking is free for the first hour.

- Tamra wrote the expression $4h$, where $h$ = the number of hours the car is in the garage.

- Pratik wrote the expression $4(h-1)$, where $h$ = the number of hours the car is in the garage.

- Luis wrote the expression $4h - 4$, where $h$ = the number of hours the car is in the garage.

Who is correct? Why?

## Student Work 1

Mr. Kropach asks his class to write an expression for the following scenario:

A parking garage charges $4 each hour, but parking is free for the first hour.

- Tamra wrote the expression $4h$, where $h$ = the number of hours the car is in the garage.
- Pratik wrote the expression $4(h-1)$, where $h$ = the number of hours the car is in the garage.
- Luis wrote the expression $4h - 4$, where $h$ = the number of hours the car is in the garage.

Who is correct?  Why?

Tamra is correct because we only know that he is paying four dollars an hour, he does get free parking for an hour but we do not know how long he stayed so the free parking is useless in this equation.

## Student Work 2

Mr. Kropach asks his class to write an expression for the following scenario:

A parking garage charges $4 each hour, but parking is free for the first hour.

- Tamra wrote the expression $4h$, where $h$ = the number of hours the car is in the garage.
- Pratik wrote the expression $4(h-1)$, where $h$ = the number of hours the car is in the garage.
- Luis wrote the expression $4h - 4$, where $h$ = the number of hours the car is in the garage.

Who is correct?  Why?

Pratik is correct. He is because he wrote 4(h-1), you'll be multiplying but subtracting 1. You are because the 1st hour is free. It represents the number of hours but you use a letter for the numbers you do not know.

### Student 3

Like Student 2, Student 3 also identifies Pratik's expression as a correct solution. However, Student 3 is unable to articulate why the expression is correct. Student 3 attempts to apply the distributive property to examine equivalent expressions, but doesn't distribute the 4 to the second term in the quantity.

### Student 4

Student 4 accurately identified Luis's expression as a correct solution. He used the test value of 5 hours to determine that the total cost would be $20, but then the first hour is free, causing a reduction of the total cost by $4. Student 4 has not considered any other ways to represent this scenario.

## USING EVIDENCE

*What would we want to ask these students? What might we do next?*

### Student 3

Why did Student 3 select Pratik's expression? We can ask him to explain how he knows that $4(h - 1)$ is correct. We will need to continue to help him develop explicit connections in his explanations. His incorrect distribution of $4(h - 1)$ may be more problematic. We need to uncover why he wrote it and could begin that effort by asking what the 1 represents in his expression $4h - 1$. Will he recognize that $4h - 1$ isn't an equivalent expression? We may find that additional work with equivalent expressions, evaluation of expressions, and associated concepts such as the distributive property is needed.

### Student 4

Student 4 has accurately explained why Luis's expression is correct. We should be sure to acknowledge how he justified his solution. His work would be useful during a classroom discussion to highlight how reasoning can be justified. Like Student 2, Student 4 doesn't recognize the equivalent expression within the task. We might challenge him to compare the values of the other expressions with his value of 5 hours so he can find his oversight.

**TASK 23A:** Mr. Kropach asks his class to write an expression for the following scenario:

A parking garage charges $4 each hour, but parking is free for the first hour.

- Tamra wrote the expression 4*h*, where *h* = the number of hours the car is in the garage.

- Pratik wrote the expression 4(*h* − 1), where *h* = the number of hours the car is in the garage.

- Luis wrote the expression 4*h* − 4, where *h* = the number of hours the car is in the garage.

Who is correct? Why?

## Student Work 3

Mr. Kropach asks his class to write an expression for the following scenario:

A parking garage charges $4 each hour, but parking is free for the first hour.

- Tamra wrote the expression 4*h*, where *h* = the number of hours the car is in the garage.
- Pratik wrote the expression 4(*h* − 1), where *h* = the number of hours the car is in the garage.
- Luis wrote the expression 4*h* − 4, where *h* = the number of hours the car is in the garage.

Who is correct? Why?

Pratik is right because 4(h-1) is correct.
4(h-1) becomes 4h-1.

## Student Work 4

Mr. Kropach asks his class to write an expression for the following scenario:

A parking garage charges $4 each hour, but parking is free for the first hour.

- Tamra wrote the expression 4*h*, where *h* = the number of hours the car is in the garage.
- Pratik wrote the expression 4(*h* − 1), where *h* = the number of hours the car is in the garage.
- Luis wrote the expression 4*h* − 4, where *h* = the number of hours the car is in the garage.

Who is correct? Why?

Luis cause if some one stays
5 hours(s=h) its $29 since the first
hours is free it's − 4 dollars

## OTHER TASKS

- What will count as evidence of understanding?

- What misconceptions might you find?

- What will you do or how will you respond?

 Visit this book's companion website at **resources.corwin.com/ minethegap/6-8** for complete, downloadable versions of all tasks.

**TASK 23B:** Write an expression for the perimeter of a rectangle whose length is three times its width.

Write an expression for the perimeter of a rectangle whose width is one third its length.

How are the expressions similar? How are they different? Why does this make sense?

**CONTENT CONNECTION**

While this task is assessing understanding of comparing expressions, it is also connected to geometry concepts of perimeter and properties of a rectangle. Extensions to this task may include comparing the area of these rectangles.

Task 23B requires students to write expressions for perimeters of rectangles with multiplicative relationships between the length and width. *For this task, students must apply their understanding of geometric relationships in order to compare the expressions.* Students may identify that the length, $l$, is equal to three times the width, $3w$, but then may not understand how to substitute in order to write an expression in terms of one variable. Students may also generate an initial correct expression, but may make errors when they simplify the expression. Some students will try to create expressions based on the same variable, whereas others will write one expression in terms of the width and the other in terms of the length. It will be important to emphasize defining variables as well as drawing pictures to represent the scenario in order to accurately compare the two rectangles.

**TASK 23C:** Antwaine is buying candy bars for his baseball team. He has a coupon to save $0.50 for every candy bar he buys. He also sees that each candy bar is on sale for 25% off.

Write an expression that models this scenario. Explain how you determined your expression.

**CONTENT CONNECTION**

Compound expressions can be challenging for students. Emphasize the need to represent the problem with one expression, and have students use their understanding of order of operations in order to accurately write the expression.

This task builds on understanding of percent problems to write an algebraic expression. Some students address the percent savings before addressing the $0.50 savings. *Students may try to write two expressions by first writing an expression for the $0.50 coupon, then writing a new expression for the 25% off.* Students who write two expressions may even use the same variable for different purposes. Defining the variable will be a critical step for students to understand how to effectively use the variable in the expression. As noted, this task and task 23B connect with other concepts. It is important that we determine why student errors occur. For example, is their error in this task a result of unfinished learning with percents or with writing expressions?

**TASK 23D:** For each expression below, write a scenario that could be used to model the expression:

$$45h + 60 \qquad\qquad 5(x - 3)$$

$$10m + 50 + 4m + 5 \qquad\qquad \tfrac{1}{2}b + 2$$

*This task is a good example of how we can increase rigor and provide for differentiation.* It affords students the opportunity to apply their understanding. It helps us determine if students understand how an expression should be used, including the order of operations to be applied for any given input value. Students may commonly try to select a scenario based on the variable, such as creating a scenario based on hours for the expression $45h + 60$. Students may be most challenged with $5(x - 3)$. They may create a scenario that would be modeled by the expression $5x - 3$. Students may also have difficulty with the fraction in the final expression. These students may write a scenario that would require an expression to divide by one-half rather than multiply by one-half. We can leverage student examples for discussion about similarity and differences.

**MODIFYING THE TASK**

**We may have to reteach or revisit single operation expressions prior to advancing to multiple operation expressions.**

NOTES

_____

_____

_____

_____

_____

_____

_____

_____

_____

_____

_____

_____

_____

_____

_____

_____

_____

_____

# BIG IDEA 24

## Evaluating Expressions

### TASK 24A

For which value of *m* will the expression $\frac{1}{2}(m-2)$ have the greatest value?

$m = 8$ $\qquad\qquad m = \frac{1}{2}$ $\qquad\qquad m = -8$ $\qquad\qquad m = -\frac{1}{2}$

**Explain your reasoning.**

## About the Task

**MODIFYING THE TASK**

**We could extend the task by asking students to consider how they might adjust the expression so that each value for *m* would create the greatest value.**

Evaluating algebraic expressions helps students develop an understanding of input and output as well as how the components of an expression work together. Predicting or anticipating the result of an expression helps students determine the reasonableness of their solutions. Applying different values to a variable enables students to explore how inputs of different magnitude affect the resulting outputs. At first glance, this may seem like a procedural task. However, the reasoning students use should help us determine whether students understand the relationship between the value of the input and the resulting output.

## Anticipating Student Responses

As we know, errors may occur from miscalculations or flawed application of order of operations. Some students may determine that the greatest value of the expression is when *m* = 8 because that is the greatest value of *m*. Students are

### PAUSE AND REFLECT

- How does this task compare to tasks I've used?
- What might my students do in this task?

 Visit this book's companion website at **resources.corwin.com/minethegap/6-8** for complete, downloadable versions of all tasks.

likely to approach the task with two specific strategies. *Some students will directly substitute each of the values of* m *into the expression to find the resulting outputs.* Students will then compare the resulting outputs. Another strategy is to consider how the expression will impact the value of each of the inputs. Some students might rule out the negative values of *m*, since these values will produce negative outputs. Students may then reason which of the positive values of *m* will produce the largest output. Some students may use the greatest value (8) or the least (−8) to find an output about which they can make a conjecture.

**MINING HAZARD**

**While this is a valid strategy, it is important to focus on the student reasoning to determine how they are developing an understanding of the relationship between input and output.**

NOTES

### Student 1

Student 1 understands that the value of $m$ should be substituted into the expression in order to evaluate the expression. Student 1 makes a number of errors as she evaluates each expression. For example, in the first expression it appears that she uses half of 8 and then subtracts 2 to get a value of 2. In the second expression, she simplifies inside the parentheses and writes that this quantity must be multiplied by $\frac{1}{2}$, but then she multiplies by 2. It is unclear how she determined her solution for the third expression. Student 1's work for the last expression shows similar errors as her work with the second expression.

### Student 2

Student 2 substitutes to compare the expressions for the given values of $m$. She correctly calculates the value for the positive values of $m$, but appears to have difficulty simplifying expressions with negative values of $m$. In one case, Student 2 multiplies a positive and a negative number and displays a positive value as the product. In another case, Student 2 calculates $-\frac{1}{2} - 2$ to be a positive $2\frac{1}{2}$. Student 2 also misuses the term *equation* in place of *expression*.

## USING EVIDENCE

*What would we want to ask these students? What might we do next?*

### Student 1

**MINING TIP**

It's helpful to isolate an idea or concept when students demonstrate a range of misconceptions or errors. We can improve understanding of one idea and reinforce it with specific practice before moving to another misconception.

We can build on Student 1's understanding of substitution. *There is a range of ideas that needs further development so that she can simplify expressions.* We might address the multiplication error in the second expression by asking what it means to take half of something. Then, we would ask her to take half of $1\frac{1}{2}$ and $-1\frac{1}{2}$. In some cases, she tried to apply the distributive property but did not distribute to each term. We might have her simplify inside the parentheses first, then have her take half of the number. It may be necessary for her to create a physical or visual model to show this representation. To do so, she might use 8 counters, subtract 2, and then take half of the counters away. It is important to have her compare accurate results with her findings so she can improve awareness of missteps.

### Student 2

We want to clarify with Student 2 the difference between an equation and an expression. She does show understanding of how to evaluate expressions, but it is clear that she needs continued work with integer operations. Using representations such as number lines can support this continued development. We can support the accuracy of her work with expressions and equations by having her evaluate and solve them before using a calculator to confirm her accuracy.

**TASK 24A:** For which value of $m$ will the expression $\frac{1}{2}(m-2)$ have the greatest value? Explain your reasoning.

$m = 8$  $\qquad\qquad m = \frac{1}{2}$  $\qquad\qquad m = -8$  $\qquad\qquad m = -\frac{1}{2}$

## Student Work 1

For which value of $m$ will the expression $\frac{1}{2}(m-2)$ have the greatest value?

① $m = 8$  $\qquad$ ② $m = \frac{1}{2}$  $\qquad$ ③ $m = -8$  $\qquad$ ④ $m = -\frac{1}{2}$

Explain your reasoning.

① $\frac{1}{2} \times (8-2) = 2$

② $\frac{1}{2} \times (\frac{1}{2} - 2) = -3$

$\qquad -1\frac{1}{2} \times \frac{1}{2} =$

③ $\frac{1}{2} \times (-8 - 2) = -5$

④ $\frac{1}{2} \times (-\frac{1}{2} - 2) = -3$

$\qquad -2\frac{1}{2} \times \frac{1}{2}$

## Student Work 2

For which value of $m$ will the expression $\frac{1}{2}(m-2)$ have the greatest value?

$\frac{1}{2}(6)$

$m = 8$  $\qquad m = \frac{1}{2}$  $\qquad m = -8$  $\qquad m = -\frac{1}{2}$

$3$  $\qquad\qquad \frac{1}{2}(-1.5)$  $\qquad \frac{1}{2}(-10)$  $\qquad \frac{1}{2}(-2.5)$

Explain your reasoning  $-.75$  $\qquad 5$  $\qquad 1.25$

I think $m = -8$ will have the greatest value because once you complete the equation, you get 5, which even if it was negative has the greatest Absolute value of all of them.

### Student 3

Student 3 correctly identifies 8 as the value of *m* that will produce the largest value for the expression. Student 3 did not use substitution to test any of the values. *She simply reasoned that because the input value was the greatest, the result would be the greatest.* It is unclear whether the student knows that this strategy may not be valid for all expressions.

### Student 4

Like Student 3, Student 4 also identifies 8 as the value of *m* that will produce the largest value for the expression. Student 4 substituted to determine what the resulting value of the expression would be. Student 4 also explains that multiplying by a fraction is the same as dividing, though the wording isn't quite accurate. She then references the other expressions, stating "the other would get a negative no matter what which means 3 would be the greatest value."

**MINING HAZARD**

Applying understanding of number sense is an important skill that tasks like this can develop. It is important for our students to clarify why their reasoning works. We also want to be on the lookout for overgeneralizing their reasoning.

*What would we want to ask these students? What might we do next?*

### Student 3

We want to acknowledge that reasoning can be much more efficient than calculating. We might offer ideas for how Student 3 can confirm her conclusions. We might note that you only need to compare the largest and smallest values of *m* rather than all four. Even with her solid reasoning in place, we should ask her if her strategy would work for all expressions or just this one. We should make sure she clearly explains why. We should also have an expression on hand, such as $-4(m - 7)$, in case she notes that the greatest value (8) will always create the greatest value of any evaluated expression.

### Student 4

Student 4 has demonstrated that she understands how to evaluate the expression for given values of *m*. She has procedure and understanding in place to help her develop more efficient reasoning similar to Student 3. In fact, providing these two students opportunities to review the other's explanation may help advance both of their understandings. Our work in these cases is to encourage students to predict outcomes with specific reasoning before finding proof of their predictions. This improves our students' ability to determine the reasonableness of their answers while advancing their reasoning and critical thinking in general.

**TASK 24A:** For which value of $m$ will the expression $\frac{1}{2}(m-2)$ have the greatest value? Explain your reasoning.

$m = 8$ $\qquad\qquad m = \frac{1}{2}$ $\qquad\qquad m = -8$ $\qquad\qquad m = -\frac{1}{2}$

### Student Work 3

For which value of $m$ will the expression $\frac{1}{2}(m-2)$ have the greatest value?

(m = 8) $\qquad\qquad m = \frac{1}{2}$ $\qquad\qquad m = -8$ $\qquad m = -\frac{1}{2}$

Explain your reasoning.

Because 8 is the biggest out of all the choices.

### Student Work 4

For which value of $m$ will the expression $\frac{1}{2}(m-2)$ have the greatest value?

(m = 8) $\qquad\qquad m = \frac{1}{2}$ $\qquad\qquad m = -8$ $\qquad m = -\frac{1}{2}$

Explain your reasoning.

this will be because m = 8 - 2 would then equal 6 and then you'll have to multiply it by $\frac{1}{2}$
$\frac{1}{2} \times 6 = 3$ because when you multiply a fraction by a whole number or a whole number by a fraction, the whole number would be divided by the fraction.
The others would get a negative no matter what which means 3 would be the greatest value.

## OTHER TASKS

- What will count as evidence of understanding?
- What misconceptions might you find?
- What will you do or how will you respond?

Visit this book's companion website at **resources.corwin.com/minethegap/6-8** for complete, downloadable versions of all tasks.

**TASK 24B:** Create two different expressions. For each expression you create, determine what the value of the variable is to produce an expression equal to 10.

**MODIFYING THE TASK**

We can modify the task to prompt students to create different expressions that meet certain criteria. For example, we might require an expression that includes integers or an expression that has two operations and a fraction.

*Task 24B is an opportunity for students to apply their understanding of evaluating expressions by generating two different expressions that are equivalent to 10.* We should look for students who use one operation in both expressions as that may be indicative of lesser confidence or comfort with evaluating expressions. We may find students who create the expressions $4 + x$ and $2 + x + 2$, $-5(-y)$ and $5(y)$, or $20 \div m$ and $40 \div m$. These ideas don't realize the intent of the task but do offer valuable insight into students' thinking. We should also look for other similarities between expressions, including the use of similar or different operations within or between expressions as well as the use of whole numbers, fractions, decimals, and/or integers.

**TASK 24C:** For each expression, create a new expression that will generate the same output when $x = 3$.

| | | | |
|---|---|---|---|
| $4x - 10$ | $-x + 4$ | $-2x + 4$ | $3x - 1$ |

This task requires students to evaluate expressions and generate a new expression that will produce the same output as the original when $x = 3$. Students need to understand how to apply inverse operations in order to generate a new expression. Students who do not have a strong understanding of inverse operations may rely on trial and error in order to find an expression that works. This task helps students develop fluencies needed to support solving equations. Some students may evaluate each expression and then create a new, rather simple expression. For example, these students might find the value of $4x - 10$ to be 2 when $x = 3$. From there, they might create the expression $x - 1$ that also produces an output of 2.

**TASK 24D:** Ansel finds that two different values for $x$ produce the same value for the expression $2x^2 - 1$. Explain why this is possible, or describe the mistake you think he is making. Create a new expression that supports your thinking.

This task provides an opportunity to begin work with expressions that include exponents, which we know to be quadratic expressions. Unlike linear expressions, quadratic expressions will have two inputs that result in the same output. Students may approach this task by testing different values of *x* to determine whether Ansel has made a valid statement. Though this is a valid strategy, it may be evidence of unfinished learning. Other students will offer that both a value and the opposite of that value produce the same value when they are squared. These students may also note the results of multiplying integers. Mathematically proficient students make and test conjectures and generalizations. This task also prompts students to create a new situation to support their reasoning, giving an opportunity to create and test their assumption.

NOTES

# BIG IDEA 25
## Equivalent Expressions

### TASK 25A

Which of the following are equivalent expressions?

| | | | |
|---|---|---|---|
| $3x - 9$ | $3(x - 3)$ | $3(x - 9)$ | $3(x - 6)$ |
| $2x + (x - 9)$ | $3(x - 4) + 3$ | $x + x + x - 9$ | $4x - (x + 9)$ |

Choose two expressions and tell how you know they are equivalent.

### About the Task

Determining equivalent expressions promotes flexibility and efficiency when evaluating expressions and solving equations. To develop fluency with identifying equivalent expressions, students must understand how the components of an expression are related. In this task, students need to consider multiple ways to express the same representation. This task provides insight into whether our students understand how to use the distributive property as well as correctly combining like terms. It also includes expressions that require multiplication and addition of integers.

### Anticipating Student Responses

There are a host of pitfalls for students with unfinished learning. Some may consider any expression that includes both $3x$ and 9 to be equivalent. These students may discount $4x - (x - 9)$ because they simply see $4x$. They may

## PAUSE AND REFLECT

- How does this task compare to tasks I've used?

- What might my students do in this task?

Visit this book's companion website at
**resources.corwin.com/minethegap/6-8**
for complete, downloadable versions of all tasks.

fail to recognize that expressions like $2x + (x - 9)$ and $4x - (x + 9)$ will also yield equivalent expressions. Others may not recognize that $x + x + x$ is $3x$. Some may carry out order of operations incorrectly. Successful students might simplify some or all of the expressions to see which are equivalent. Other students will look for common features of the expressions to narrow their scope of comparison. *Even successful students may only look for one or two equivalent expressions, rather than considering each and every expression as a potential viable equivalent expression.*

**MINING HAZARD**

**We must determine if students with accurate but incomplete solutions understand the mathematics.**

NOTES

### Student 1

Student 1 correctly identified three equivalent expressions. Yet, she doesn't recognize that there were other equivalent expressions. She focused on expressions in which 3 was distributed. She overlooked any expression that required combining like terms other than the basic $x + x + x - 9$ expression.

### Student 2

Like Student 1, Student 2 correctly applies the distributive property and recognizes some equivalent expressions by combining like terms. Student 2 is also able to see that $2x + (x - 9)$ will produce an equivalent expression. She fails to recognize that $3(x - 4) + 3$ and $4x - (x + 9)$ produce equivalent expressions. Interestingly, each of these examples requires multiplication and addition with negative integers.

*What would we want to ask these students? What might we do next?*

### Student 1

We can build on Student 1's ability to correctly apply the distributive property. She also is able to recognize that $x + x + x$ is equivalent to $3x$. We should ask her how she was able to eliminate the other expressions. Her response will give insight into the misconceptions she may have. She may be able to autocorrect as she takes a closer look at other expressions. Or, it may reveal an inability to combine like terms with more complicated expressions. If the latter is the case, we may want to encourage her to substitute and evaluate to create cognitive dissonance and purpose for further instruction.

### Student 2

Student 2 applies the distributive property with positive integers and recognizes some equivalent expressions by combining like terms. Our first steps are to determine if she can distribute negative integers and combine like terms with negative integers. Understanding this will help us decide if she needs additional support with integer operation or if she simply overlooked these possible solutions.

**TASK 25A:** Which of the following are equivalent expressions?

$3x - 9$        $3(x - 3)$        $3(x - 9)$        $3(x - 6)$

$2x + (x - 9)$        $3(x - 4) + 3$        $x + x + x - 9$        $4x - (x + 9)$

Choose two expressions and tell how you know they are equivalent.

### Student Work 1

Which of the following are equivalent expressions?

$(3x - 9)$        $(3(x - 3))$        $3(x - 9)$        $3(x - 6)$

$2x + (x - 9)$        $3(x - 4) + 3$        $(x + x + x - 9)$        $4x - (x + 9)$

Choose two expressions and tell how you know they are equivalent.

3x-9 is equivalent to 3(x-3).
You need to distribute 3(x-3).
3x-9 is the simplest form of
3(x-3).

### Student Work 2

Which of the following are equivalent expressions?

$(3x - 9)$        $3(x-3)$        $3(x - 9)$        $3(x - 6)$

$(2x + (x - 9))$        $3(x - 4) + 3$        $x + x + x - 9$        $4x - (x + 9)$

3x-9

Choose two expressions and tell how you know they are equivalent.

$x + x + x - 9$

$3x-9$

$3(x+3)$

$3x-9$

### Student 3

Student 3 applies the distributive property with positive integers. She is also able to simplify the expression $3x - 12 + 3$ to verify that this is another equivalent expression. However, she overlooks expressions that have $2x$ or $4x$ as the leading term.

### Student 4

Student 4 accurately selects some correct expressions but omits others and selects incorrect responses as well. Her selections show evidence of ability to combine like terms with positive terms. But she doesn't show mastery of the distributive property. She incorrectly selected $3(x - 9)$ and appeared to consider $3(x - 6)$ (note light erasing) as potential solutions. This shows that she is not sure how to address the second term inside the parentheses.

## USING EVIDENCE

*What would we want to ask these students? What might we do next?*

### Student 3

We can build on Student 3's ability to distribute with positive integers and recognition of some equivalent expressions by combining like terms. Student 3 seems able to multiply integers and combine like terms by adding $-12$ and $+3$. We should ask the student why $2x + (x - 9)$ and $4x - (x + 9)$ are not equivalent expressions. Her explanation may reveal unfinished learning relative to combining like terms in all cases.

### Student 4

Student 4 shows that she can combine like terms in some situations such as $2x + (x - 9)$ and $x + x + x - 9$. She doesn't show full understanding of the distributive property. It is especially noted as the two top left expressions are examples of it. For Student 4 and others, we might use algebra tiles to model and compare the expressions. It will be critical that similarities and differences between the models are highlighted. Then, connections must be made between the physical/visual models and the symbolic expressions. With understanding in place, engaging activities such as memory style games and "I have, Who has" can be used to practice and reinforce.

**TASK 25A:** Which of the following are equivalent expressions?

| | | | |
|---|---|---|---|
| $3x - 9$ | $3(x - 3)$ | $3(x - 9)$ | $3(x - 6)$ |
| $2x + (x - 9)$ | $3(x - 4) + 3$ | $x + x + x - 9$ | $4x - (x + 9)$ |

Choose two expressions and tell how you know they are equivalent.

### Student Work 3

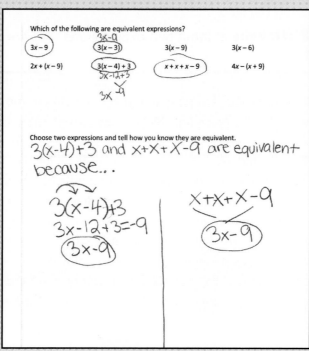

### Student Work 4

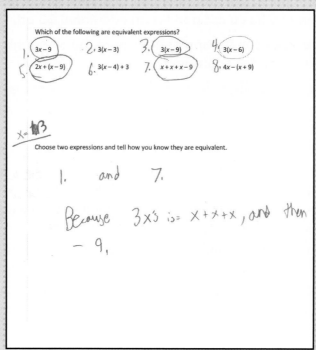

# OTHER TASKS

- What will count as evidence of understanding?
- What misconceptions might you find?
- What will you do or how will you respond?

 Visit this book's companion website at **resources.corwin.com/ minethegap/6-8** for complete, downloadable versions of all tasks.

**TASK 25B:** Write at least two expressions that are equivalent to 4(2*m* – 5).

Creating equivalent expressions is more rigorous than identifying them. In this task, students are asked to apply their understanding of the distributive property to generate equivalent expressions. Most students will apply the distributive property to produce the expression 8*m* – 20. Because the students need to create at least two expressions, students have a number of options for creating another expression. Students may choose to decompose the quantities by producing expressions like 4*m* + 4*m* – 20. Others may show an intermediate step for the distributive property to produce the expression 4(2*m*) + 4(–5). Students who are challenged to apply the distributive property may need additional practice or may need us to reteach the concept. We may consider doing this with decomposed whole number factors before reintroducing variables. For example, we could connect these ideas with 4 × 14 as 4(10 + 4). Or we may examine 7 × 19 as 7(10 + 9) or 7(20 – 1).

**TASK 25C:** Mary and Julie are holiday shopping. Julie found a pair of shoes that is discounted $5 and then an additional 20% is taken off of the discounted price. Mary found a pair of shoes that is 80% of the original price, then she has a coupon for an additional $5 off.

Write an expression to represent each of the discounts and determine if the expressions are equivalent.

**CONTENT CONNECTION**

Connecting expressions and percents can help develop deeper understanding of both concepts.

*Representing real-world contexts with algebraic expressions can support our understanding and reasoning with the situations.* Some students may be quick to judge both situations and expressions to be the same because both feature a $5 and 20% discount. They may even create incorrect expressions to justify their "logic." Some students may understand the situations, noting that the two are not equal. Yet, even students who can reason about the problem may be unable to write the expressions $(p - 5)(1 - 0.20)$. Students will need to pay particular attention to the language "off of the discounted price" versus "80% of the original price." This nuance can cause students to inaccurately represent the scenario. Once students generate each expression, they must then determine whether these expressions are equivalent. They must apply understanding of equivalent expressions to determine whether $(1 - 0.20)(p - 5)$ is equivalent to $0.80p - 5$. Some students will use a random, original price to test their expressions.

**TASK 25D:** Clark thinks that expressions can be equivalent even if they have different numbers. His examples are

$6p + 16$

$2(3p + 8)$

$10p - 2(2p + 5) + 6$

$7p - p + 17 - 1$

**Bill thinks that expressions can be equivalent even if they have different variables. His examples are**

$10x + 12$

$10z + 12$

$10s + 12$

$10y + 12$

**Who do you agree with? Why?**

Task 25D provides an opportunity to address two misconceptions students may have. One misconception that students hold is that expressions must look alike to be equivalent. These students will agree with Bill. Others who agree with Bill will do so because they don't believe equivalent expressions can have different numbers or operations. *This flawed logic may even be defended with a value being provided for each variable in Bill's expressions.* As we know, these different variables can have different values that would counter this flawed logic. Students who agree with Clark may overlook the possibility that Bill's expressions could have equivalent values. Students who agree with Clark will combine terms and apply order of operations to make their case. This task is a good opportunity for groups of students to have conversations about their perspectives and justifications.

**MINING TIP**

Students' misconceptions are not random. Their misconceptions are grounded in beliefs that they hold and arguments that they can make.

NOTES

# BIG IDEA 26

## Writing Equations

### TASK 26A

Your mom's muffin recipe will produce 8 muffins. The recipe calls for $2\frac{1}{2}$ cups of milk. Your mom plans to bake muffins for the school bake sale. She sees that she has 15 cups of milk in the refrigerator.

Write an equation that will allow you to solve for the number of muffins she can make.

Use pictures, numbers, or words to explain how you found your equation.

**MINING HAZARD**

**As students begin work with problem solving and equation writing, we should be sure to make use of realistic, friendly values so that their focus is not distracted by complicated computations.**

### About the Task

Writing equations is a natural extension of writing expressions. Equations are a tool for representing and solving problems. Writing equations may help students "see" and reason about the mathematics within a problem. This task asks students to create an equation to find a solution to the problem. They might decide to create an initial equation to find the necessary ingredients for the next equation. *The task moves beyond values of whole numbers, yet the fraction/mixed number remains manageable.*

### Anticipating Student Responses

This task provides flexibility to represent the equation in different ways. Some students may first address the available 15 cups of milk and the $2\frac{1}{2}$ cups of milk needed per recipe. Multiplication with fractions may be problematic,

## PAUSE AND REFLECT

- How does this task compare to tasks I've used?

- What might my students do in this task?

Visit this book's companion website at
**resources.corwin.com/minethegap/6-8**
for complete, downloadable versions of all tasks.

causing calculation issues or adjustments. Because of this or preference, students may choose to rewrite $2\frac{1}{2}$ as 2.5. Students may find the quotient and then multiply the result by 8 muffins per recipe. Others may note that there are 2 groups of $2\frac{1}{2}$ in 5 so there are 6 in 15 because 15 is 3 times greater than 5. Students may try to write two equations rather than finding a way to represent the entire scenario as one equation. Students may also write an expression rather than an equation. We should be careful to monitor how students are defining their variable and using it in the context of the problem.

NOTES

### Student 1

Student 1 defines $x$ as "how many muffins she can make." He writes the equation $15 \div 2\frac{1}{2} = x$. Student 1 understands that the 15 cups of milk must be divided by the cups of milk required for each batch of muffins. However, Student 1 doesn't account for the 8 muffins the recipe produces. Because of this, he writes an equation to determine how many batches of muffins his mom can make rather than the total number of muffins.

### Student 2

Student 2 defines $M$ as the number of muffins. He writes the equation $15 \div 2.5 \div 8 = M$. Student 2 describes that he should divide 15 by 2.5 to see how much milk you will have. He also understands that he must account for the 8 muffins, but mistakenly thinks that he must divide the batches by 8 rather than multiply by 8.

## USING EVIDENCE

*What would we want to ask these students? What might we do next?*

### Student 1

**MINING TIP**

**Positioning students to find and make sense of their errors is more powerful than correcting their errors for them.**

We should first be sure that Student 1 understands what the problem asks of him. It may be a misunderstanding, though unlikely, because his explanation notes that $x$ represents the number of muffins she can make. *We might have Student 1 calculate the quotient and explain what the solution means in the context of the situation to see if he can make sense of the error.* He should find a quotient of 6, which we can ask him to evaluate in the context of the problem. It can be helpful for students to identify what is known about problems before creating an equation. For this problem, we can note that a batch creates 8 muffins, but we don't know how many batches can be made.

### Student 2

Student 2's error is centered around how to address the 8 muffins made per batch. It may be a challenge specifically connected to creating and solving equations as he believes he will need to "get to" $8M$. We can have him solve the equation to find a value of $M$, resulting in a solution of 0.75 muffins. It's possible that this solution alone will cause him to reconsider his approach because less than 1 muffin in a batch doesn't make sense. We might have Student 2 write an equation for each step of the problem. This would create an equation for the number of batches and the total number of muffins made.

**TASK 26A:** Your mom's muffin recipe will produce 8 muffins. The recipe calls for $2\frac{1}{2}$ cups of milk. Your mom plans to bake muffins for the school bake sale. She sees that she has 15 cups of milk in the refrigerator. Write an equation that will allow you to solve for the number of muffins she can make. Use pictures, numbers, or words to explain how you found your equation.

## Student Work 1

Your mom's muffin recipe will produce 8 muffins. The recipe calls for 2 ½ cups of milk. Your mom plans to bake muffins for the school bake sale. She sees that she has 15 cups of milk in the refrigerator.

Write an equation that will allow you to solve for the number of muffins she can make.

$8 = 2\frac{1}{2}$          $15 \div x =$

$15 \div 2\frac{1}{2} = x$

Use pictures, numbers, or words to explain how you found your equation.

I used "x" for my answer. X is meaning how many muffins she can make.

## Student Work 2

Your mom's muffin recipe will produce 8 muffins. The recipe calls for 2 ½ cups of milk. Your mom plans to bake muffins for the school bake sale. She sees that she has 15 cups of milk in the refrigerator.

m = muffins

Write an equation that will allow you to solve for the number of muffins she can make.

$15 \div 2.5 \div 8 = m$

Use pictures, numbers, or words to explain how you found your equation.

first, I thought I should divide 15 and 2.5 because you need to see how much milk you'll have to make a lot of muffins.

### Student 3

Student 3 wrote an inaccurate equation: $15 \div 2\frac{1}{2} = 6 \times 8 = m$. He then draws 15 cups of milk, some of which have a line down the center, insinuating a half. He multiplies this by 8, noting that is equal to $m$. His work shows that he understands to divide the 15 cups of milk by the cups needed for each recipe. Student 3 also shows that he must multiply this amount by the 8 muffins made per batch.

### Student 4

Student 4 writes the equation $2\frac{1}{2} \times 8m = 15$. He draws a pictorial representation and uses proportional reasoning to show that $2\frac{1}{2}$ cups of milk will make 8 muffins, so 5 cups of milk will make 16 muffins, and 15 cups of milk will make 48 muffins. It appears he originally defined $x$ as the number of muffins, but then chose to use the variable $m$, which he has not defined.

*What would we want to ask these students? What might we do next?*

**MINING HAZARD**

**Requiring students to explain their ideas in words may undermine or limit their ability to express their thinking. Student 3 is an example of sound mathematics reasoning represented with nothing more than pictures and symbols.**

### Student 3

Student 3 shows understanding of the problem and how the parts of the problem are used to find a solution. *Without a single word, he explains his approach, though a more clear connection to the equation is desirable.* Student 3's most pressing error is within the equation. Our work is to revisit the meaning of an equation. Our first step may be with single-step equations before moving to problems and equations with two steps. It is essential that he understands that an equation represents two equivalent values. In this problem, the number of muffins that can be made (unknown) is equivalent to 8 times the number of batches that can be made ($15 \div 2\frac{1}{2}$).

### Student 4

Student 4 has applied understanding of proportional relationships to set up and solve an equation to find the number of muffins that can be made. While many students will write an equation of the form $(15 \div 2\frac{1}{2}) \cdot 8 = m$, Student 4 has determined a correct equation where the variable is not isolated. Our next steps with him are to remind him to define the variable and communicate how his equation connects to his representation. Student 4's work should be part of our classroom discussion as it can support his classmates' development of flexible thinking relative to problem solving and writing equations.

## Student Work 3

Your mom's muffin recipe will produce 8 muffins. The recipe calls for 2 ½ cups of milk. Your mom plans to bake muffins for the school bake sale. She sees that she has 15 cups of milk in the refrigerator.

Write an equation that will allow you to solve for the number of muffins she can make.

$$15 \div 2\tfrac{1}{2} = 6 \times 8 = m$$

Use pictures, numbers, or words to explain how you found your equation.

## Student Work 4

Your mom's muffin recipe will produce 8 muffins. The recipe calls for 2 ½ cups of milk. Your mom plans to bake muffins for the school bake sale. She sees that she has 15 cups of milk in the refrigerator.

Write an equation that will allow you to solve for the number of muffins she can make.

$$2\tfrac{1}{2} \times 8m = 15$$

Use pictures, numbers, or words to explain how you found your equation.

You have 15 milk.

## OTHER TASKS

- What will count as evidence of understanding?
- What misconceptions might you find?
- What will you do or how will you respond?

 Visit this book's companion website at **resources.corwin.com/minethegap/6-8** for complete, downloadable versions of all tasks.

**TASK 26B:** Bonnie has $18 to spend at the grocery store. She must buy 2 gallons of milk, which sell for $3.50 a gallon, and 5 yogurts, which cost $0.79 each. She also wants to buy some snacks. Write an equation that will allow Bonnie to determine how much money she has available to spend on snacks.

 **CONTENT CONNECTION**

**This task builds on understanding of writing equations and makes a connection to linear functions explored in the study of functions. This type of task helps build an understanding of how the slope and y-intercept impact different representations of a linear function.**

Task 26B is similar to task 26A in that it helps us determine if students can accurately write an equation to represent a real-world solution. Proficient students may represent the unknown in two different ways. One possible representation would be to have the money available for snacks, $m$, remain on the same side of the equation as the other purchases, such as $2(3.50) + 5(0.79) + m = 18.00$. Students may also choose to isolate the money available for snacks. To do this accurately, students must subtract each of the other expenses from $18.00 to represent the equation as $m = 18.00 - 2(3.50) - 5(0.79)$. We can support students who are developing their understanding with questions that direct them to identify what is needed to buy and the quantities of those things. We might ask them to identify how they can find the costs of those items and how they might be combined. Before working with values, we might have students create a representation of the components of the problem—milk, yogurt, snacks, and available money.

 **MINING TIP**

**We should take note of the contexts that students use for creating and solving problems. It may provide insight into limited understanding of application. It may also highlight the situations and contexts we predominantly feature during instruction.**

**TASK 26C:** Write a scenario that may be modeled by the equation $y = -2x + 12$. Be sure to identify what $x$ and $y$ represent in your scenario. Describe what the −2 represents in the context of your scenario.

In this task, students need to develop a scenario that can be modeled by an equation with two variables. *This task serves as a bridge to writing equations for linear functions.* Students not only have to find a linear situation, but also must consider how a negative rate of change impacts the scenario. Some students will choose a scenario based on time, such as number of hours, days, or weeks. *Proficient students will recognize that the −2 will represent a decrease by 2 for each unit defined as the independent variable, such as "decreasing by $2.00 each week."*

**TASK 26D:** Mr. Branch asks the class to write an equation for the perimeter of a rectangle with a width of 4 centimeters and a perimeter of 32 centimeters.

**Write two different equations that would model this scenario. Explain why there can be two different equations for this scenario.**

Equations can model many mathematical concepts. Task 26D provides students the opportunity to represent the perimeter of a rectangle using two different equations. Students need to apply their understanding of equivalent expressions to generate two equivalent equations to represent the scenario. Proficient students will clearly define variables as they build an equation, like $2(4) + 2l = 32$, where $l$ represents the length of the rectangle. Students may choose to apply a variety of properties of equality to write other equations, such as $8 + 2l = 32$, $2l + 8 = 32$, $32 = 8 + 2l$, or $2(4 + l) = 32$. Yet, connecting equations to other mathematics concepts can prove to be challenging for students and for our ability to determine what we need to help them better understand. This task is a good example of just that. For this task, we must determine if unsuccessful students are challenged by the ideas of perimeter or the ideas of writing equations. Our next instructional steps should then be based on those specific concepts.

NOTES

# BIG IDEA 27
## Solving Equations

### TASK 27A

Samantha solves the equation $4(2y - 7) = 64 + 12y$. Her work is shown below:

$$4(2y - 7) = 64 + 12y$$

$$2y - 7 = 16 + 3y$$

$$5y = 23$$

$$y = 4.6$$

Review Samantha's work to determine if she has found the correct solution. If she has made any mistakes, identify the mistake.

### About the Task

There are steps to solving equations. Even so, those steps are built on the fundamental understanding that the two expressions are equivalent. To solve equations, our students need to understand how to balance an equation, simplify expressions, and assess reasonableness of solutions. This task provides students with an opportunity to apply their understanding of solving equations to assess another student's work. The prompt does not indicate whether the solution is correct, so the student must verify whether each step is correct.

### Anticipating Student Responses

Most students will try to follow each step to determine if the work is correct. *Some students may try to solve the equation with an alternate approach to*

## PAUSE AND REFLECT

- How does this task compare to tasks I've used?

- What might my students do in this task?

 Visit this book's companion website at **resources.corwin.com/minethegap/6-8** for complete, downloadable versions of all tasks.

determine whether they find the same value for y. Some students may also decide to verify the solution by substituting the answer into the original equation to determine if the equation is true. Others may note the work Samantha did between each new line. We may find that some of our students assume that dividing both sides by 4 is incorrect, stating that Samantha should have applied the distributive property first. We should also anticipate miscalculations that lead to disagreements.

**MINING HAZARD**

It is important for students to understand that a correct solution could be generated using incorrect reasoning. If a student does not analyze the strategy used, he or she may not be distinguishing between correct and incorrect methods for solving an equation.

NOTES

### Student 1

Student 1 incorrectly states that there is an error in the first step. She holds the misconception that the only way to solve this problem is to apply the distributive property. Student 1 notices that Samantha simplified the right side of the equation to 16 + 3$y$, but admits that she does not know how this was generated. Student 1 also makes a computation error when she simplifies –28 – 64 and provides a result of –36.

### Student 2

Like Student 1, Student 2 also incorrectly states that "she (Samantha) didn't multiply at the beginning." This student overlooks the right side of the equation becoming 16 + 3$y$. Student 2 also makes errors when solving the equation herself. She subtracts 8 from 8$y$ on one side and when subtracting 8 from 64 on the other side. She makes another error as she subtracts 28 from both sides of the equation, resulting in an expression rather than an equation.

*What would we want to ask these students? What might we do next?*

### Students 1 and 2

Because Student 1 and Student 2 have similar misconceptions, we can work with them in similar ways. First, we would want to address the errors that they are making as they solve the equation. One approach is to have them substitute their solution into the original equation to verify that they have found a correct solution. When these students determine that they have made a mistake, allow them to analyze their own work. Student 2, in particular, may need physical or visual models for support as she solves. Algebra tiles may be used to help students distinguish between variable terms and constants. Physical models are helpful in other ways. For example, they may help students see 4 groups of 2$y$ – 7. It may be desirable to have each student review the work of the other.

Students can continue to evaluate their solution by testing the original equation and each new line as they solve it. They might even apply this approach to the lines of Samantha's work. They should find that the error is between the second and third line. We should draw attention to what the mistake appears to be and determine why there is no mistake between lines 1 and 2 of Samantha's work.

## Student Work 1

Samantha solves the equation $4(2y - 7) = 64 + 12y$. Her work is shown below:

$$4(2y - 7) = 64 + 12y$$

$$2y - 7 = 16 + 3y$$

$$5y = 23$$

$$y = 4.6$$

Review Samantha's work to determine if she has found the correct solution. If she has made any mistakes, identify the mistake.

$$4(2y - 7) = 64 \cdot 12y$$

$$8y - 28 = 64 + 12y$$
$$-8y \qquad\qquad -8y$$

$$-28 = 64 + 4y$$
$$-64 \quad -64$$

$$\frac{-36}{4} = \frac{4y}{4}$$

$$-9 = y$$

Samantha did not multiply $4 \cdot 2$ or $4 \cdot -7$. Also, I do not know where she got the 16 and 3y.

## Student Work 2

Samantha solves the equation $4(2y - 7) = 64 + 12y$. Her work is shown below:

$$4(2y - 7) = 64 + 12y$$

$$2y - 7 = 16 + 3y$$

$$5y = 23$$

$$y = 4.6$$

Review Samantha's work to determine if she has found the correct solution. If she has made any mistakes, identify the mistake.

$$4(2y - 7) = 64 + 12y$$

$$8y - 28 = 64 + 12y$$
$$-8 \qquad\qquad -8$$

$$28 = 56 + 12y$$
$$-28 \quad -28$$

$$\frac{28}{12} + \frac{12y}{12}$$

$$y = 2.33$$

Samantha's mistake was at the beginning. She didn't multiply from the start.

### Student 3

As with others, Student 3 also incorrectly states that Samantha failed to apply the distributive property. Unlike Students 1 and 2, Student 3 does not make any errors as she solves the equation. Student 3 is a good example of students who find correct answers but may not have the depth of understanding we desire. She is procedurally accurate, but her oversight may indicate flawed conceptual understanding. Student 3 finds the actual mistake in Samantha's work by stating that "$3y - 2y$ is not $5y$."

### Student 4

Student 4 also applies the distributive property to solve. She generates the correct solution of $y = -23$. She identifies that the solution in the prompt is incorrect. But, she has not made any specific reference to what the mathematical mistake is. It is unclear if Student 4 recognizes that there *is* a mistake or if she assumes there is one because her solution ($y = -23$) isn't the same ($y = 4.6$).

## USING EVIDENCE

*What would we want to ask these students? What might we do next?*

### Student 3

Student 3 understands how to use the distributive property and assumes that this must be applied first to solve an equation. She believes this is one of the errors she recognizes in Samantha's work. We must work with Student 3 to recognize that the first two lines in Samantha's work are equivalent. We might make use of substitution, which may cause some dissonance for Student 3. Physical models may be needed for her to see the relationship between the two equations. Essentially, we want her to establish that there can be more than one way to begin to solve an equation. We can reinforce this notion by providing additional practice and discussion with similar equations.

### Student 4

We should have Student 4 identify what specific mistake(s) Samantha made. This will provide insight as to whether she has a misconception about the distributive property or whether she has simply overlooked the mistake of subtracting $3y$. Based on her response, we may approach her similarly to Student 3. If it is an oversight, we should continue to help Student 4 develop clear and complete explanations. One approach may be to have Student 4 compare her writing to a benchmark example of a previous or fictional student. We should be sure that benchmark samples highlight different features of clarity. For this task, we might share a sample that uses a substitution strategy to identify an error or a sample that connects correct steps adjacently to the errors. It is important to keep in mind that lengthy, complex writing isn't needed for a good explanation.

## Student Work 3

Samantha solves the equation $4(2y - 7) = 64 + 12y$. Her work is shown below:

$$4(2y - 7) = 64 + 12y$$

$$2y - 7 = 16 + 3y$$

$$5y = 23$$

$$y = 4.6$$

$4(2y-7)=64+12y$

$8y-28=64+12y$

$\phantom{8y}+28\phantom{=9}+28$

$\phantom{8}8y\phantom{+}=92\phantom{+}+12y$

$\phantom{8y=9}-12y\phantom{+}-12y$

$\phantom{8y}\frac{-4y}{-4}=\frac{92}{-4}$

Review Samantha's work to determine if she has found the correct solution. If she has made any mistakes, identify the mistake.

-she didn't muliply correctly at the first distribution

-she got 3y out of nowhere

-3y -2y is not 5y

-Answer is incorrect.

## Student Work 4

Samantha solves the equation $4(2y - 7) = 64 + 12y$. Her work is shown below:

$$4(2y - 7) = 64 + 12y$$

$$2y - 7 = 16 + 3y$$

$$5y = 23$$

$$y = 4.6$$

Review Samantha's work to determine if she has found the correct solution. If she has made any mistakes, identify the mistake.

this solution is not correct

$4(2y-7)=64+12y$

$8y-28=64+12y$

$-8y\phantom{-28=64+}-8y$

$\overline{\phantom{xxxxxxxxxxxxx}}$

$-28\phantom{xx}=64+4y$

$-64\phantom{x=}-64$

$\frac{-92}{4}\phantom{xx}\frac{4y}{4}$

$-23=y$

## OTHER TASKS

- **What will count as evidence of understanding?**
- **What misconceptions might you find?**
- **What will you do or how will you respond?**

 Visit this book's companion website at **resources.corwin.com/ minethegap/6-8** for complete, downloadable versions of all tasks.

**TASK 27B:** Mrs. Dellota is planning a birthday party for her daughter at the skating rink. She must pay $9.00 per child for food and $3.50 per child for skate rentals. She has a budget of $150.00. Write and solve an equation to determine how many children she may have attend the party.

Equations can be applied to all sorts of problems and situations. Multi-step situations, such as that in this task, are quite challenging for students to represent with equations. It can also be challenging for us to decipher where our students went wrong and what to do next. Students may not initially identify that the cost per child is $12.50, leading to an equation such as $9.00c + 3.50c = 150.00$, where $c$ represents the number of children attending the party. Other students may attempt to subtract similar values ($9.00c$ and $3.50c$) from $150.00. We may find that some of our students solve the problem (12 students) but are unable to write an equation to do so. We have students who create an equation, solve it, and find the value of $c$ to be 120 due to a calculation error. Each of these examples is cause for different teacher moves. For some, we will have to continue work with writing equations. Similar actions may be taken with those who solve the problem but are unable to write an equation. Those who miscalculate signal a need for discussion about the reasonableness of their solutions.

**TASK 27C:** Examine each of the equations.

   **Equation 1:** $3x + 4 = -5x + 8$

   **Equation 2:** $8x = 12$

   **Equation 3:** $6x + 8 = -10x + 16$

   **Equation 4:** $3x = 4$

   **Equation 5:** $3x + 4 = -10x + 16$

**Some of the equations above have the same solution. Determine which equations will generate the same solution. Explain your reasoning.**

**MINING TIP**

Tasks such as this are great opportunities for students to exchange observations, insights, strategies, and claims. The exchange of these ideas nurtures the developing reasoning of our students.

We cannot factor out the importance of reasoning when working with and solving equations. *That reasoning is grounded in seeing relationships between terms and expressions within equations.* This type of task serves as a necessary precursor to building a conceptual understanding of solving systems of equations, particularly using linear combinations. We may find that some of our students are compelled to solve each equation to prove which have the same solution. This is a viable strategy, but it isn't efficient. Some students will recognize that equation 3 is equation 1 multiplied by a constant factor of 2. Some students will make inaccurate assumptions by seeing similarities in the equations. For example, equations 2 and 4 may be a distraction because 3 × 4 is 12, so there appears a relationship between the equations. Others may assume that equations 3 and 5 will generate the same solution because both terms in the left expression have been multiplied by a factor of 2 and the right expressions are unchanged.

## TASK 27D: Max wonders if $2(2x + 3) = 12$ is the same as $2x + 3 = 6$. Use numbers or models to explain if they are or are not the same.

Our students' work with the distributive property begins as early as third grade. They extend the idea to multiplying multi-digit numbers, decimals, and even mixed numbers. But their ability to compute accurately does not necessarily indicate deep understanding of it. This task offers an opportunity for students to demonstrate a conceptual understanding for using algebraic properties to solve an equation. This task is an opportunity to examine their understanding. Some students will model how to apply the distributive property to generate the equation $4x + 6 = 12$. Other students will recognize that this equation represents 2 groups of $2x + 3$, which is equivalent to 12, and will then reason that 1 group of $2x + 3$ is equal to 6. Students who solve both equations to justify that the two equations are the same show understanding but necessarily an understanding of the distributive property.

NOTES

# BIG IDEA 28
## Inequalities

### TASK 28A

> You are planning a family trip to a family amusement park. For all roller coasters, you must be at least 54 inches to ride. For water rides, you must be at least 48 inches to ride. For some of the kiddie rides, you must be between 36 and 52 inches to ride.
>
> Represent the height requirements of each ride with inequalities and represent each on a number line.
>
> Is it possible to ride all rides in the park? If so, who may ride all of the rides?

## About the Task

Inequalities help us consider conditions that make a situation true. They are useful in all sorts of contexts. This task provides a real-world context for students to represent compound inequalities. Our students are often asked to represent and graph an inequality, but typically students are asked to represent a single inequality or there is no context for the inequality. Students must determine a strategy for representing the height requirements with inequalities and then determine if there are cases in which some children would be able to ride all of the rides. The authentic nature of this task should help many students make sense of the context, the number line representation, and the accuracy of their solutions. By representing the inequalities on a number line, students will be able to see that no child can ride all of the rides, but children between 48 and 52 inches can ride all rides except for roller coasters.

## PAUSE AND REFLECT

- How does this task compare to tasks I've used?

- What might my students do in this task?

Visit this book's companion website at
**resources.corwin.com/minethegap/6-8**
for complete, downloadable versions of all tasks.

## Anticipating Student Responses

Students may encounter difficulties determining a strategy for representing and comparing multiple inequalities. A common error we might anticipate is a failure to accurately define the variable. Students may choose different variables to represent the same quantity due to the condition or situation. For example, they might think of the roller coaster as $r > 54$; water rides as $w > 48$; and kiddie rides as $36 < k < 52$. In each of these cases, the variable is used to represent a height. This error will affect students' ability to accurately compare the inequalities. Or, students might compare inequalities without recognizing the flaw of different variables. *Some students will be able to accurately represent inequalities with one variable ($h > 54$, $h > 48$, $36 < h < 52$). Other students may create number line representations that show overlapping solutions.*

**CONTENT CONNECTION**

The study of inequalities provides a natural connection to using number lines explored in the number system. Students have the opportunity to extend understanding of number lines to algebraic inequalities.

NOTES

## Student 1

Student 1 represents each of the height requirements on a different number line. He shades the correct region for each number line. However, he doesn't answer whether it is possible to ride all of the rides at the park. Because of this, it is unclear if he is able to compare the graphs in the context of the amusement park scenario.

## Student 2

Like Student 1, Student 2 creates three accurate representations of the height restrictions on three different number lines. He correctly states that it is not possible for someone to ride all of the rides. Student 2 arranges the three number lines vertically in an attempt to compare. However, he uses a different scale for each line, which likely leads to his faulty comparison.

## USING EVIDENCE

*What would we want to ask these students? What might we do next?*

## Student 1

Our questions might help Student 1 make sense of the prompt. We might ask if it is possible to ride all of the rides in the park. This will help to determine if he can compare the three height restrictions. If he is unable to respond correctly to the question, we might offer approaches to making comparisons less challenging. We can acknowledge the use of number lines but suggest that he create them with the same scale or locate each directly above the others. We may also consider challenging him to make the comparisons on the same number line. As with other students, we may consider having him test heights of different people to check which rides they may be able to ride.

## Student 2

Like Student 1, we can build on Student 2's understanding. We might ask him how he can use the graphs to verify that no one can ride all of the rides. We should be sure to highlight that his arrangement of the number lines is a good approach. We should also ask him why the number lines each start in a similar location but show different values. It may be helpful to ask if someone can ride two of the rides to establish the reasoning necessary for satisfying the prompt. In fact, he and Student 1 may benefit most by comparing inequalities of two situations before extending their understanding to three or more comparisons.

**TASK 28A:** You are planning a family trip to a family amusement park. For all roller coasters, you must be at least 54 inches to ride. For water rides, you must be at least 48 inches to ride. For some of the kiddie rides, you must be between 36 and 52 inches to ride. Represent the height requirements of each ride with inequalities and represent each on a number line. Is it possible to ride all rides in the park? If so, who may ride all of the rides?

## Student Work 1

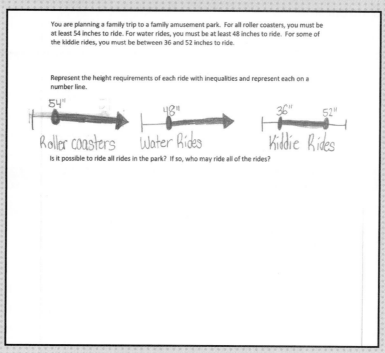

You are planning a family trip to a family amusement park. For all roller coasters, you must be at least 54 inches to ride. For water rides, you must be at least 48 inches to ride. For some of the kiddie rides, you must be between 36 and 52 inches to ride.

Represent the height requirements of each ride with inequalities and represent each on a number line.

54"  Roller coasters    48"  Water Rides    36"  52"  Kiddie Rides

Is it possible to ride all rides in the park? If so, who may ride all of the rides?

## Student Work 2

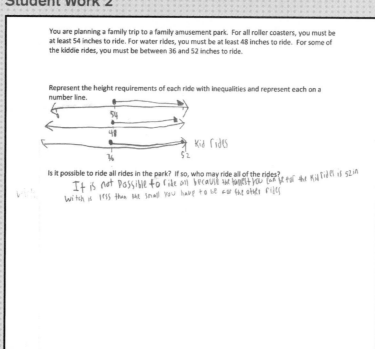

You are planning a family trip to a family amusement park. For all roller coasters, you must be at least 54 inches to ride. For water rides, you must be at least 48 inches to ride. For some of the kiddie rides, you must be between 36 and 52 inches to ride.

Represent the height requirements of each ride with inequalities and represent each on a number line.

54

48

36    52    Kid rides

Is it possible to ride all rides in the park? If so, who may ride all of the rides?
It is not Possible to ride all because the longest you can be for the kid rides is 52 in witch is less than the small you have to be for the other rides

### Student 3

Student 3 attempts to plot each of the inequalities on a single number line. He uses a "jump" strategy on the number line, which may indicate that only whole numbers are part of the solution set. Student 3 makes the incorrect claim that "people 54 inches and taller can ride them all." This signals that Student 3 is unable to interpret the number line representation that he created.

### Student 4

Like Student 3, Student 4 also plots each of the inequalities on one number line. Student 4 realizes that there is a gap between 52 and 54 inches, so he decides to provide a creative response, "You would have to be 51 but wear heels on one ride and take them off." He then comments that without the support of the heels, you cannot ride all of the rides.

*What would we want to ask these students? What might we do next?*

### Student 3

There are two misconceptions we need to address. We should be sure that he understands that the height restrictions are inclusive of all heights within the interval, not just whole number heights. An easy way to check would be to ask him to consider a child who has a height of 40.5 inches. The second misconception to address with Student 3 is how to interpret multiple height restrictions. We might have him draw each inequality separately and adjacently. This should help him distinguish between each of the inequalities. We should ask about where the inequalities overlap and what that would tell us about the rider.

### Student 4

Student 4 has a clever insight to make up for mathematics that doesn't seem right to him. He does understand how to show inequalities and how the gap in them influences his answer. Like the others, he may benefit from using more than one number line before transitioning back to a single representation. It may be a good idea for Student 4 to compare and draw conclusions about two of the riders. His diagram may also be problematic. We might encourage him to use color or labels so that he can better "see" the relationships. It might also be a good idea for Student 3 and Student 4 to compare and contrast their representations.

**TASK 28A:** You are planning a family trip to a family amusement park. For all roller coasters, you must be at least 54 inches to ride. For water rides, you must be at least 48 inches to ride. For some of the kiddie rides, you must be between 36 and 52 inches to ride. Represent the height requirements of each ride with inequalities and represent each on a number line. Is it possible to ride all rides in the park? If so, who may ride all of the rides?

### Student Work 3

You are planning a family trip to a family amusement park. For all roller coasters, you must be at least 54 inches to ride. For water rides, you must be at least 48 inches to ride. For some of the kiddie rides, you must be between 36 and 52 inches to ride.

Represent the height requirements of each ride with inequalities and represent each on a number line.

Is it possible to ride all rides in the park? If so, who may ride all of the rides?

Yes, people 54 inches and taller can ride them all.

### Student Work 4

You are planning a family trip to a family amusement park. For all roller coasters, you must be at least 54 inches to ride. For water rides, you must be at least 48 inches to ride. For some of the kiddie rides, you must be between 36 and 52 inches to ride.

Represent the height requirements of each ride with inequalities and represent each on a number line.

Is it possible to ride all rides in the park? If so, who may ride all of the rides?

you would have to be 56 but where heels, on one ride take then off

If you have no heels you cant be because they do not all overlap

# OTHER TASKS

- **What will count as evidence of understanding?**
- **What misconceptions might you find?**
- **What will you do or how will you respond?**

Visit this book's companion website at **resources.corwin.com/ minethegap/6-8** for complete, downloadable versions of all tasks.

**TASK 28B:** An elevator can hold a maximum of 1500 pounds. If the average person weighs 160 pounds, write and solve an inequality to determine how many people could ride in the elevator at the same time. Explain your reasoning.

Many of us have wondered if an elevator will hold everyone safely as more and more people enter the car with us. Task 28B offers an authentic context for students to apply their understanding of inequalities. Students must write and solve an inequality to determine the maximum number of people who can ride on an elevator at the same time. Students may use the wrong inequality symbol by confusing maximum and minimum. Students may also fail to assess the reasonableness of their solution by providing a decimal solution for the number of people. Students may also round up their solution to the nearest whole number, which will provide an answer that exceeds the elevator capacity. In each of these cases, it is important for us to ask questions to help students make sense of the context and their solution.

**TASK 28C:** A school bus can hold 42 students. Ms. Zuckerman is planning a field trip for the 310 students in the school. How many school buses does Ms. Zuckerman need to order? Justify your reasoning.

Ms. Zuckerman forgot to include 22 teachers. She thinks she might need another bus. Do you agree or disagree that she needs another bus? Use pictures, numbers, or words to explain your thinking.

**MINING HAZARD**

Using visual models is a valid strategy to apply. We should then ask follow-up questions that require students to connect their representations to an inequality.

Task 28C is another real-world context for writing and solving inequalities. After solving, students must adjust their solution to account for 22 additional people to ride the bus. Students may approach this problem in a few ways, including to choose not to write an inequality. It is important that students understand that the problem represents the minimum number of buses that must be ordered to hold the passengers. *Some students may represent the problem with a tape diagram or similar visual model by creating groups of 42.* If students choose this strategy, they may simply continue to make groups with the 22 additional teachers.

**TASK 28D:** Which of the following inequalities have solutions with values greater than 5?

$$5 < x \qquad\qquad x - 3 > 2 \qquad\qquad x + 4 < 9$$

$$3x - 4 < 2x + 1 \qquad -2x < -10 \qquad -x - 7 < 2$$

**Choose one inequality above and explain how you know that the solution is greater than 5.**

**Write two new inequalities in which the solutions have values greater than 5.**

This task requires students to evaluate inequalities in different forms to determine whether the solution to the inequality is greater than 5. Some may calculate to find their solutions. But their explanations may lack a description of how they reasoned to find their solution. Each of the incorrect inequalities represents common errors. For example, $-2x < -10$ highlights the misunderstanding students have when dividing with a negative. They do not understand that dividing by a negative impacts the inequality. The inequality $x + 4 < 9$ highlights the misconception of students who try to solve inequalities as equations. Students need to understand that solutions to inequalities have multiple solutions, whereas an equation has one solution. With these misconceptions in mind, any incorrect identification will provide insight into the misconception(s) students have. Extension of the task may offer insight into what students understand and/or what they are comfortable with relative to inequalities. For example, single-operation inequalities may indicate that we should focus on inequalities with more than one operation. Students who create inequalities that neglect integers may be signaling that we need more work with inequalities and these features.

NOTES

# BIG IDEA 29
## Function Tables

### TASK 29A

Describe how a rule for a function table works.

Write a rule that would have the following function table:

| Input (x) | Output (y) |
|-----------|------------|
| 1 | 5 |
| 2 | 7 |
| 5 | 13 |
| 8 | 19 |

## About the Task

Function tables are useful for seeing relationships between quantities and for solving problems. Work with function tables provides students with an introduction to developing an understanding of functional relationships. Often, work with function tables is relegated to finding the output for designated inputs by applying a given rule. Occasionally, students may be asked to work in reverse by finding an input for a specific output. Yet in this task, the function table is complete. Students will work to find and describe the relationship between the inputs and outputs.

## Anticipating Student Responses

Our students' approach to this task will shed light on their understanding of functional relationships. Some students may hold the misconception that

### PAUSE AND REFLECT

- How does this task compare to tasks I've used?
- What might my students do in this task?

Visit this book's companion website at
resources.corwin.com/minethegap/6-8
for complete, downloadable versions of all tasks.

the vertical values are related and may create one or even two incorrect rules or functions. Other students may look for either an explicit relationship or a recursive relationship. For students looking for a recursive relationship, they may try to identify outputs for consecutive inputs, then describe the relationship as "adding two." Students looking for an explicit relationship will try to generate a formula that will produce the output for the given input. Some may begin multiplying the inputs by a number and then comparing to the output table. For example, if a student multiplies each input by 2, she may notice that the output is 3 less than output in the function table. Students may make errors representing the function rule with subtraction rather than addition. Other students may neglect to see that the input table is not increasing by consecutive numbers.

NOTES

### Student 1

Student 1 demonstrates some understanding of what a function is as she describes, "every input can only have one output." She also states, "It's what you do to $x$ to get $y$" to describe the purpose of how a rule for a function table works. However, she is unable to generate a rule for the function table. She incorrectly provides the equation $y = x + 5$. There is no indication how she generated this rule, though it could be applied to the second row. In fact, that may be exactly how she found her rule, indicating that her description may not be as valid as we first thought. Be cautious, though; inference can be dangerous when interpreting student work. It is imperative that we rely on the evidence. Her writing implies that she understands a rule, but coupled with her incorrect rule may indicate the notion that each row has its own rule or function.

### Student 2

Student 2 describes that "$y$ would be the one that is constantly going up in a pattern and $x$ would go up with it by 1 but it could have left over." She also reverses the slope and $y$-intercept for her equation by writing the function rule $y = 3x + 2$. She understands that the $y$-intercept should be 3 but doesn't transfer this understanding to her function rule.

*What would we want to ask these students? What might we do next?*

### Student 1

We should acknowledge that Student 1 has some understanding of what functions are, though we should be sure that she has full understanding. In other words, does she understand the function to be what is done to every input rather than a single input? We may easily uncover this understanding or misconception by asking her to justify the rule she provides. Or, we might ask her if the table has another rule. Though we may not be able to establish strategies for finding the rule in this task, we can have her consider if her rule is valid. To do this, we can simply have her apply it to each row. This may lead her to pose different possibilities. As she does, we may also look to see if each new generation has a single operation and what that operation is.

### Student 2

Student 2 creates a rule with more than one operation. Like Student 1, her rule only works with one of the rows. She attempts to generate an output for a new input using her rule but doesn't apply it to known inputs and outputs. We will need to clarify her understanding of how a function rule works. As with Student 1, we can ask her to substitute the given inputs to find the associated outputs. Both students will benefit from additional work with functions. We should consider using real-world contexts to help them relate situations to the function table. As they work with these new tasks, we should spotlight different strategies for attempting to find specific rules.

**TASK 29A:** Describe how a rule for a function table works. Write a rule that would have the following function table:

| Input (x) | Output (y) |
|-----------|------------|
| 1 | 5 |
| 2 | 7 |
| 5 | 13 |
| 8 | 19 |

## Student Work 1

Describe how a rule for a function table works.

Every input can only have one output.
Its what you do to x to get y.

Write a rule that would have the following function table:

| Input (x) | Output (y) |
|-----------|------------|
| 1 | 5 |
| 2 | 7 |
| 5 | 13 |
| 8 | 19 |

Rule: $y = x + 5$

## Student Work 2

Describe how a rule for a function table works.

y would be the one that is constantly going up in a pattern and x would go up with it by 1 but it could have left over

Write a rule that would have the following function table:

| Input (x) | Output (y) |
|-----------|------------|
| 1 | 5 |
| 2 | 7 |
| 5 | 13 |
| 8 | 19 |

$y = 3x + 2$

when x is at 0 y is at 3 because it goes down by 2 each time or up by 2.

### Student 3

At first glance, it may seem that Student 3 has a valid strategy, yet there are significant limitations to the strategy she uses. Student 3 uses a strategy that is only appropriate for linear models. Note that as students begin to work with linear functions, they begin to overgeneralize and assume that all function equations may be generated with a common difference and $y$-intercept. It is important to expose students to different function types as a reminder that not all functions may be linear. Student 3 describes that this equation works "because when $x$ is 0 $y$ is 3 and the common difference is 2." It appears that Student 3 only considers the first two input and output values, which happen to have consecutive inputs. Student 3's description of a function rule is also based solely on a linear function, rather than a generalized description for any function table.

### Student 4

Student 4's description of a function rule is based on the idea of manipulating an input to generate an output. Her statement, "the rule is how much it goes up by in the process" needs some clarification. Student 4 correctly generates a function rule $y = 2x + 3$ to model the table of values. She notices that each input could be added by a number that is increasing by 1 (e.g., $1 + 4 = 5$, $2 + 5 = 7$, $3 + 6 = 9$, etc.). It is not clear how she translates this pattern into the function rule.

*What would we want to ask these students? What might we do next?*

### Student 3

Student 3 finds the correct solution with flawed understanding. It is critically important that we address the misconception that all function tables are linear. We can ask her how she verified that the function rule works for all of the inputs. She will likely use substitution to confirm her results. We can furnish a new table of values for a linear function with no consecutive inputs and observe how she works to find the function. We might present a third function table that represents the equation $y = x^2$. As she shows success with each new table, we can help her refine her description of how a function works.

### Student 4

We need Student 4 to clarify her explanation for a function rule, and ask how her rule shows "how much it goes up by in the process." She may hold a misconception that the values of inputs are always increased for corresponding outputs. If she holds this misconception, our immediate step is to offer a table with a rule that decreases. We want to know how the pattern translates to the equation $y = 2x + 3$. If her explanation is flawed or she is unable to explain, we can have her show how she could verify that her equation is correct. Then we can introduce another linear function table to ensure that she can apply her strategy to other functions.

**TASK 29A:** Describe how a rule for a function table works. Write a rule that would have the following function table:

| Input (x) | Output (y) |
|---|---|
| 1 | 5 |
| 2 | 7 |
| 5 | 13 |
| 8 | 19 |

## Student Work 3

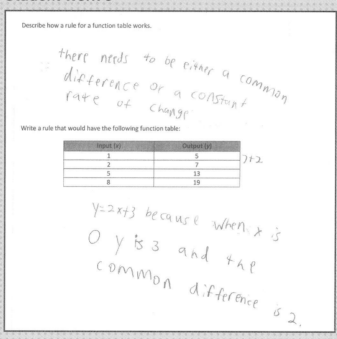

Describe how a rule for a function table works.

there needs to be either a common difference or a constant rate of change

Write a rule that would have the following function table:

| Input (x) | Output (y) | |
|---|---|---|
| 1 | 5 | )+2 |
| 2 | 7 | |
| 5 | 13 | |
| 8 | 19 | |

y=2x+3 because when x is 0 y is 3 and the common difference is 2.

## Student Work 4

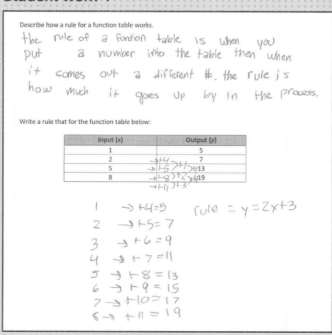

Describe how a rule for a function table works.

the rule of a funtion table is when you put a number into the table then when it comes out a different #. the rule is how much it goes up by in the process.

Write a rule that for the function table below:

| Input (x) | Output (y) |
|---|---|
| 1 | 5 |
| 2 | →+4 →+1→ 7 |
| 5 | →+5 →+1 +13 |
| 8 | →+8 +2 +19 |
| | →+11 )+3 |

1 →+4=5        rule = y=2x+3
2 →+5=7
3 →+6=9
4 →+7=11
5 →+8=13
6 →+9=15
7 →+10=17
8 →+11=19

Visit this book's companion website at **resources.corwin.com/ minethegap/6-8** for complete, downloadable versions of all tasks.

**TASK 29B:** Which of the following tables of values do not represent functional relationships? Explain your reasoning.

**A.**

| Input (x) | Output (y) |
|:---:|:---:|
| 1 | 2 |
| 2 | 2 |
| 5 | 2 |
| 8 | 2 |

**B.**

| Input (x) | Output (y) |
|:---:|:---:|
| 1 | 3 |
| 2 | 7 |
| 1 | 9 |
| 2 | 11 |

**C.**

| Input (x) | Output (y) |
|:---:|:---:|
| 1 | 15 |
| 2 | 13 |
| 3 | 11 |
| 4 | 9 |

**D.**

| Input (x) | Output (y) |
|:---:|:---:|
| 9 | 1 |
| 8 | 2 |
| 7 | 1 |
| 6 | 2 |

**MODIFYING THE TASK**

We could extend this task or modify it by asking students to simply create a table of values that represents a functional relationship and one that does not.

*Task 29B requires students to apply their understanding of a function by using the known values to determine if there is a functional relationship.* Some students may note that all tables are functions because none of the values are duplicated in any row. A related misconception is the unsuccessful student who selects A or D because he or she sees a value in the columns that is repeated. These students relate that they don't understand that it is possible for different inputs to produce the same output value. For example, Table A might have a rule similar to $O(x)$ + 2. Some students may verify their solutions by plotting the points to see how these tables are represented graphically. Some students may quickly recognize and explain that a function may not have an input that produces different output values, so they will discount Table B.

**TASK 29C:** Use the functions to find the missing values for each table. Explain how you found the missing input and outputs for each table.

$y = 4x + 1$

| Input ($x$) | Output ($y$) |
|---|---|
| 1 | |
| | 13 |
| 11 | |
| | 1 |

$y = (x - 1) \div 4$

| Input ($x$) | Output ($y$) |
|---|---|
| | 5 |
| | 13 |
| 11 | |
| 0 | |

**MODIFYING THE TASK**

We can modify the function in this task to feature one operation for students who are still developing understanding of function tables or those who are still developing computational fluency. We can also create a nonlinear example to extend students' thinking.

*Task 29C is intentionally designed to see if our students understand the relationship between the function equation and the function table.* The functions in the table are inverses of one another. By providing different input and output values, we will be able to see if students are able to use the correct order of operations as well as the inverse operations in order to find the missing table values. We might expect some of our students to substitute values and solve the equation to find the related unknown. Some students may see the relationship between either the functions or the values within the tables after completing one of the tables. Students should be able to explain how the function can be used to find either the input or the output.

**TASK 29D:** For each function table, describe the rule. Then determine if the function is a linear function. Explain your reasoning.

| Yards | Feet |
|---|---|
| 1 | 3 |
| 2 | 6 |
| 5 | 15 |
| 8 | 24 |

| Units Sold | Employee Sales Bonus |
|---|---|
| 7 | 1 |
| 14 | 2 |
| 32 | 5 |
| 39 | 6 |

**MINING TIP**

Real-world contexts such as conversions or familiar unit rates can help students develop understanding of functions and function tables.

**MINING HAZARD**

Students offer correct answers for the wrong reasons. The second table does not represent a perfect linear relationship but would most likely be modeled with a linear function.

*This task provides students with an opportunity to examine two tables of values with real-world contexts so that they can determine if the relationships are functional.* Students must then analyze the relationships further to determine if either is linear. Proficient students may be able to connect the yards-to-feet example to their understanding of unit rate as slope explored in ratios and proportional relationships. *Some students may incorrectly state that the employee sales bonus table of values represents a linear relationship because the table of values is increasing.* For these students, we want to help them understand how the idea of constant rate of change connects to the table of values for linear relationships. We also want students to translate these tables of values to a graph as another means to confirm or challenge their conjectures.

# BIG IDEA 30

## Reasoning About Graphing

### TASK 30A

The graph represents the linear function $y = mx + b$. Sketch the graph of $y = mx + b + 2$.

Explain how you found your solution.

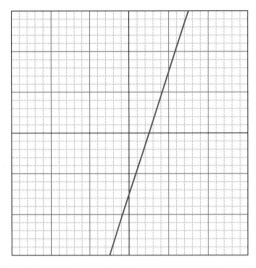

**CONTENT CONNECTION**

This task serves to connect understanding between functional relationships and geometric transformations explored in geometry. Students need to build an understanding that, like a closed figure, a function can be transformed in the coordinate plane.

## About the Task

Graphing equations helps us understand them. It enables us to make predictions, draw conclusions, and solve problems. Early work with linear functions helps students build an understanding of functional behavior and how graphs are transformed by components of the function. *In this task, students inspect an equation of a linear function as a graphical transformation.* Exploring functions

### PAUSE AND REFLECT

- How does this task compare to tasks I've used?

- What might my students do in this task?

 Visit this book's companion website at **resources.corwin.com/minethegap/6-8** for complete, downloadable versions of all tasks.

positions students to make connections between transformations and the graph of the transformed function. In this task, students consider a vertical translation.

## Anticipating Student Responses

Our students are likely to approach this problem in one of three ways. Some students will first use the graph to create a table of values. *Students will then add 2 to each of the outputs. These students will then plot points that contain the new outputs.* Some students may find the equation of the sample line displayed and add two to the equation and graph the resulting function. Others will recognize that the transformation results in a line that is shifted up two units. Others may confuse the transformation and reason that the line shifts two points to the right.

**MINING TIP**

Correct solutions can be process oriented. Plotting points from a table of values to generate a graph shows a process perspective. Though valid, reliance on procedure may not be indicative of full understanding and reasoning. We want students to understand functions as inputs and outputs, and as objects in the plane that may be transformed, known as object perspective. Both are critical to developing an understanding of functions.

NOTES

## Student 1

Student 1 draws a line with a different slope and $y$-intercept. He doesn't understand that the function $y = mx + b + 2$ results in vertical translation or that the line should be similar in general. Instead, he assigns arbitrary values for the slope and $y$-intercept and then adds 2 to the equation. His new equation is graphed correctly, signaling no consideration that this equation was not related to the original graph.

## Student 2

Student 2 understands that adding 2 to a function will result in a translation in the plane. He incorrectly assumes that this will result in a horizontal translation and graphs a line that is translated 2 units to the right of the original line.

*What would we want to ask these students? What might we do next?*

## Student 1

Student 1 shows that he can graph lines and solve equations. We ask him to compare the two lines as well as the two equations. We might ask how the new linear graph relates to the original equation. We might ask, "How does the original graph relate to your function $y = 1x + 1$?" We can introduce slightly new functions for him to graph, such as $y = 1x - 1$ and $y = 1x + 2$. We can have him compare these as well as possible others. In each new situation, we want him to compare and contrast equations and lines to draw *his* conclusions. As he develops understanding, we can begin to ask him to compare equations and predict how the lines will be related before he graphs.

## Student 2

We should acknowledge that Student 2 is attempting to reason about the mathematics. His reasoning is rooted in thinking that positive change moves to the right. Because he understands that adding 2 to an equation will represent a translation, we can have him test his belief. We can have him substitute the same values for both equations recording the results in a table. Before doing so, we should confirm that if his conjecture is true, the $y$-values for the translation should be 2 greater than the $y$-values for the original equation. When he finds that this is not true, ask him how else the line may be translated and how he might establish that it has been translated that different way.

### Student Work 1

The graph below represents the linear function $y = mx + b$. Sketch the graph of $y = mx + b + 2$.

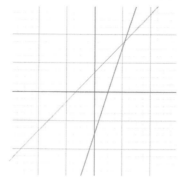

Explain how you found your solution.

$m = 1$          $y = 1x + 1 + 2 =$

$b = 1$          $y = 1x + 3$

Since mx equals 1x and b+2
equals

### Student Work 2

The graph below represents the linear function $y = mx + b$. Sketch the graph of $y = mx + b + 2$.

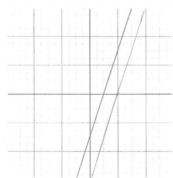

Explain how you found your solution.

The original function was Y=mx+b and
the other one is Y=mx+b+2, the 2 isn't the
Y, the Slope or the Y-intercept so I decided to
Sketch the line 2 spaces to the right
because the function says add(+) 2.

### Student 3

Student 3 correctly graphs a vertical translation of the line by 2 units. To do this, Student 3 determines an equation for the line, then adds 2 to find the equation $y = 3x + -6 + 2$. He then graphs this new equation to find the new line. Student 3 demonstrates a process perspective by finding the equation of the line and then producing a new equation in order to graph the results.

### Student 4

Student 4 correctly graphs a vertical translation of the line by 2 units. He explains that "To get the second graph, all I had to do was make sure that the $y$-coordinate was 2 more than the original graph." Student 4 demonstrates an object perspective as he reasons about the effect of adding 2 to the equation.

*What would we want to ask these students? What might we do next?*

### Student 3

First, we must acknowledge that Student 3 uses a valid strategy for finding a graph of the function $y = mx + b + 2$. Our next step is to help him develop an object perspective for greater understanding and efficiency. He conveys that he understands the different ideas needed for this. We can have him compare the differences between the two linear graphs connecting those differences to the equations. We can have him create new lines and then graph the function $y = mx + b + 2$ before discussing the results. We might have him discuss observations of translated lines and predict how the equations of the two are related. A traditional matching activity with cards showing equations of lines and cards noting translation results may be another way for him to develop his object perspective. As with most ideas in mathematics, discussion with others will better develop his understanding as he is exposed to different strategies, insights, and perspectives.

**MINING TIP**

We should be cautious about having students continuously justify their reasoning with procedure. It is wise to do so occasionally, but reliance on this approach may undermine our students' desire to reason or their confidence.

### Student 4

*Because Student 4 demonstrates an object perspective, we can extend his thinking to other types of transformations.* Before doing so, we might have him relay how he could confirm that his translated graph is accurate. We can have this student explore other transformations, such as $y = m(x + 3) + b$, $y = mx + b - 5$, and $y = m(x + 7) + b$. In each new situation, we can have him predict how the original graph may be affected by these types of transformations before testing. Doing this will help him develop his own observational skills, reasoning, and understanding.

### Student Work 3

The graph below represents the linear function $y = mx + b$. Sketch the graph of $y = mx + \underline{b + 2}$.

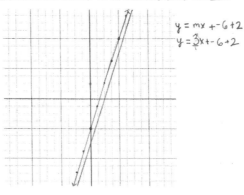

$$y = mx + {}^-6 + 2$$
$$y = 3x + {}^-6 + 2$$

Explain how you found your solution.

I found my solution by replacing m and b with numbers. The two equations are the same thing, just one is adding two. It adds two to the y-intercept, so you just go up 2 for the y-intercept and use the same slope.

### Student Work 4

The graph below represents the linear function $y = mx + b$. Sketch the graph of $y = mx + b + 2$.

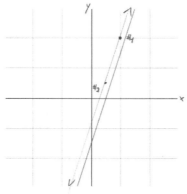

Explain how you found your solution.

To get the second graph all I had to do was make sure that the y coordinate was 2 more than the original graph.

OTHER TASKS

- What will count as evidence of understanding?

- What misconceptions might you find?

- What will you do or how will you respond?

 Visit this book's companion website at **resources.corwin.com/ minethegap/6-8** for complete, downloadable versions of all tasks.

**TASK 30B:** Write a scenario that could be modeled with the following graph.

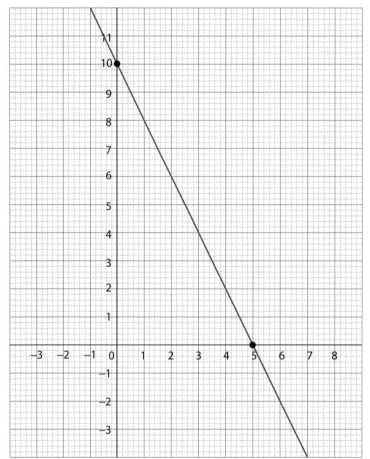

Creating graphical representations of real-world contexts enables us to solve problems, make conjectures, and look for patterns. Task 30B requires our students to do just that. Students will need to recognize that the scenario must be based on a constant decrease. Common student errors include descriptions like "the quantity decreases as the temperature decreases." This error shows that both variables are decreasing, which would result in a linear function with a positive slope. Students must recognize that the dependent variable must decrease at a constant rate, while the independent variable increases at a constant rate. Real-world examples might be the volume of water in a bathtub after the drain has been opened, the descent of a landing airplane, or the battery life of a laptop computer.

**TASK 30C:** Jason is driving to his grandparents' house. They live 40 miles away. He is driving 50 miles per hour. Sketch a graph to represent the remaining distance Jason has to travel based on how long he has been driving.

Task 30C is another example of a graphical representation of a real-world scenario. The challenging component of the task is that students must represent the remaining distance rather than the distance traveled or completed. Students may incorrectly graph a line with positive slope to represent the distance traveled based on time traveled, rather than a line with negative slope to show the decreasing distance remaining as he continues to drive. This and other tasks that require students to represent real-world contexts can be quite challenging. They are good opportunities for students to work with partners or groups to find solutions before sharing ideas with the entire group. We can help students make sense of these problems by asking them questions that can promote metacognitive thinking. Even the simplest of questions have great value in these situations. For this task, questions like, "Will he have farther to travel the longer he drives?" or "Can you predict what the graph should look like and why?" are good examples of the types of questions we want our students to eventually ask themselves.

**TASK 30D:** How do the graphs of the functions $y = 2x$ and $y = \frac{1}{2}x + 3$ compare?

This task provides us with the chance to assess how students are able to conceptualize and describe differences between two linear functions. Our students should be able to recognize that both functions represent linear functions with positive slope. Students need to also make distinctions between rates of change and $y$-intercepts. *Students may create tables of values or graphs so that they are able to make comparisons.* Proficient students will be able to use the information provided in the equations to generate accurate comparisons. We can develop fluency with these ideas as we build fluency with other mathematics concepts. We should be sure that students understand the concepts and then have ample opportunities to explore and discuss the relationships between the functions and their graphs. It may also be wise to consistently have students predict and explain the shape of a graph of a function before finding the actual results. We can then ask students to compare their predictions with their findings to help them improve what *they* should anticipate mathematically.

 **MINING HAZARD**

While producing different representations helps students to compare functions, this strategy also may show a lack of understanding of how the equation itself sheds light on functional behavior.

NOTES

# BIG IDEA 31
## Comparing Functions

### TASK 31A

Three different linear functions below are represented in three different ways:

Function A:

$y = 2x + 3$

Function B:

| x | y |
|---|---|
| 1 | 9 |
| 2 | 7 |
| 3 | 5 |
| 4 | 3 |

Function C:

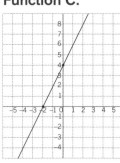

Which function has the greatest rate of change? Does any pair of functions have the same rate of change? Justify your answers.

## About the Task

Representing mathematics concepts in varied ways is evidence of deep understanding of those concepts. Representations of functions can help us see relationships, make conjectures, and solve problems. Comparing functions through representations is another essential skill that requires students to identify key features of functions. Though our students have likely compared functions, it's also likely that they have done so where each function is

### PAUSE AND REFLECT

- How does this task compare to tasks I've used?

- What might my students do in this task?

Visit this book's companion website at **resources.corwin.com/minethegap/6-8** for complete, downloadable versions of all tasks.

represented in the same way. The representations of functions in this task provide insight about the function's behavior. The task provides an opportunity to make sense of the representations to compare rates of change.

## Anticipating Student Responses

There are different approaches to finding a solution to this task. We might anticipate that some of our students will be challenged by one or more of the representations of these functions. For some, they may have worked with each in isolation and so they are challenged to compare the functions. Some of our students will try to convert each function into the same type of representation in order to compare. For example, students may graph functions A and B in order to compare them to function C. Students may also choose to write an equation for each function in order to compare. Some students will recognize that function B is a linear function with negative slope. They will also identify that functions A and C both have a slope of positive 2. Even those who can work within different representations may make an error like comparing *y*-intercepts rather than comparing slopes.

NOTES

### Student 1

Student 1 only identifies function C as the function with the greatest rate of change. She doesn't recognize that functions A and C each have the same rate of change. Her justification shows some significant flaws in thinking. She states, "Function C because it has a positive and negative rate of change and none of them are going at the same rate." Student 1 is confusing rate of change with the function being positive (above the *x*-axis) and negative (below the *x*-axis).

### Student 2

Student 2 correctly finds an equation to represent function B. However, she calculates the slope of function C incorrectly. Student 2 uses the slope formula but reverses the point $(-2, 0)$ as $(0, -2)$ in the formula to generate a slope of $-8$ rather than 2. Student 2 doesn't appear to understand how the negative slope impacts the comparison of slopes. She shares that 8 indicates the greatest slope, which contradicts her solution of $-8$.

## USING EVIDENCE

**MINING TIP**

**Students may hold understanding of a concept through one representation. We should first ensure that understanding is in place for a specific representation before prompting with different representations.**

*What would we want to ask these students? What might we do next?*

### Student 1

Student 1's misconceptions must be addressed. We need to revisit the concept of rate of change to establish that it addresses how the output changes as the input changes. We can ask Student 1 to describe "slope" and explain her understanding as the rate of change. We can investigate how to determine the greatest slope from each representation. She can then compare the slope of the three. We need to confirm that Student 1 recognizes that functions A and C have the same rate of change. *It may be wise to examine her ideas relative to one representation.*

### Student 2

We can have Student 2 find the slope of line C using the graph rather than the slope formula. We might also redirect her to the error made by using the slope formula. With revised solutions, she can then compare the three slopes. Also with corrections, Student 2 should recognize that the slopes of functions A and C are the same. We should also inspect her comparison of negative slope and positive slope as it may signal additional unfinished learning. Student 2 is good evidence of the need for students to discuss their approaches and solutions. When doing so, many will find their flawed approaches by considering the disagreement between their ideas and the ideas of their classmates.

**TASK 31A:** Compare the three functions represented in three different ways. Which function has the greatest rate of change? Does any pair of functions have the same rate of change? Justify your answers.

## Student Work 1

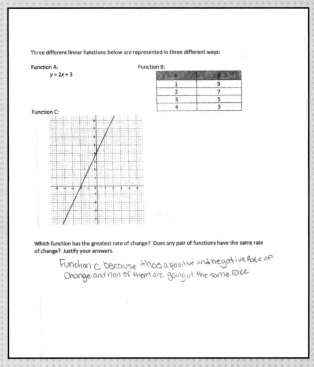

Three different linear functions below are represented in three different ways:

Function A:
$y = 2x + 3$

Function B:

| x | y |
|---|---|
| 1 | 9 |
| 2 | 7 |
| 3 | 5 |
| 4 | 3 |

Function C:

Which function has the greatest rate of change? Does any pair of functions have the same rate of change? Justify your answers.

Function C because it has a positive and negative rate of change and non of them are going at the same rate

## Student Work 2

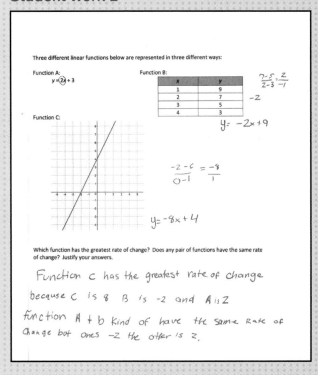

Three different linear functions below are represented in three different ways:

Function A:
$y = 2x + 3$

Function B:

| x | y |
|---|---|
| 1 | 9 |
| 2 | 7 |
| 3 | 5 |
| 4 | 3 |

$\frac{7-5}{2-3} = \frac{2}{-1}$

$-2$

$y = -2x + 9$

Function C:

$\frac{-2-6}{0-1} = \frac{-8}{1}$

$y = -8x + 4$

Which function has the greatest rate of change? Does any pair of functions have the same rate of change? Justify your answers.

Function C has the greatest rate of change because C is 8 B is -2 and A is 2
function A + b kind of have the same Rate of change but ones -2 the other is 2.

### Student 3

Student 3 incorrectly identifies function B as the function with the greatest rate of change. She correctly determines that the slope of function B is –2 but makes an error finding the y-intercept. She also incorrectly identifies the slope of function C to be –2 rather than 2. Student 3's justification doesn't provide much clarity about her ideas. It's possible that she compared the three y-intercepts rather than the slopes.

### Student 4

Student 4 incorrectly states that "all three linear functions have the same rate of change." This is because she makes a calculation error with function B, generating a slope of 2 rather than –2. She then concludes, "Functions A, B, and C have a rate of change of 2." Student 4 correctly finds the rate of change for functions A and C.

*What would we want to ask these students? What might we do next?*

### Student 3

Student 3 needs additional work finding slope, considering it as the rate of change, and considering how to compare it as evidenced by her errors. It is important to confirm that she did compare the y-intercepts rather than the slopes and why that is inaccurate. We should determine if she can articulate what slope is before more work is done to develop strategies and procedures for finding it. As understanding of and ability to find slope takes shape, we can mingle different representations to extend her understanding. As with Student 1, slope and rate of change must be explicitly connected. We can then revisit tasks such as this and others so that she can apply her refined understanding.

### Student 4

Incorrect answers are not always indicative of misconceptions or misunderstandings. Student 4 provides evidence that she understands how to find and compare the rates of change for the three functions. Her calculation leads her to an incorrect solution. We might have her describe what is happening to the y-values for function B as the x-values increase. It may be helpful to have her sketch a graph of these points to see that function B is decreasing. This might help her see that some sort of error has occurred. Student 4 also reminds us of the importance of thinking critically about our results. It would be interesting to see if she would draw the same conclusion if each linear function was represented the same way.

**TASK 31A:** Compare the three functions represented in three different ways. Which function has the greatest rate of change? Does any pair of functions have the same rate of change? Justify your answers.

## Student Work 3

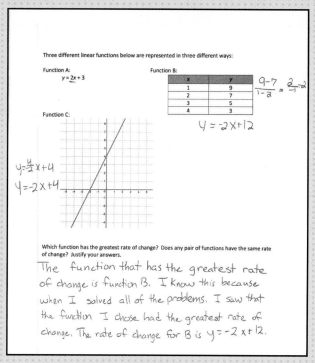

Three different linear functions below are represented in three different ways:

Function A:
$y = 2x + 3$

Function B:

| x | y |
|---|---|
| 1 | 9 |
| 2 | 7 |
| 3 | 5 |
| 4 | 3 |

$\frac{9-7}{1-2} = \frac{2}{-1} = -2$

$y = -2x + 12$

Function C:

$y = \frac{4}{2}x + 4$

$y = -2x + 4$

Which function has the greatest rate of change? Does any pair of functions have the same rate of change? Justify your answers.

The function that has the greatest rate of change is function B. I know this because when I solved all of the problems. I saw that the function I chose had the greatest rate of change. The rate of change for B is $y = -2x + 12$.

## Student Work 4

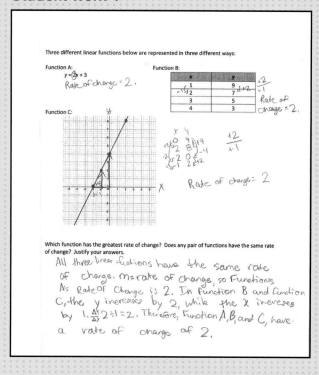

Three different linear functions below are represented in three different ways:

Function A:
$y = 2x + 3$
Rate of change = 2.

Function B:

| x | y |
|---|---|
| 1 | 9 |
| 2 | 7 |
| 3 | 5 |
| 4 | 3 |

Rate of change = 2.

Function C:

| X | Y |
|---|---|
| 0 | 4 |
| 2 | 8 |
| 7 | 0 |
| 2 | 2 |

Rate of change = 2

Which function has the greatest rate of change? Does any pair of functions have the same rate of change? Justify your answers.

All three linear functions have the same rate of change, m=rate of change, so Functions As Rate of Change is 2. In function B and function C, the y increases by 2, while the x increases by 1. $\frac{\Delta y}{\Delta x}$ 2÷1=2. Therefore, Function A, B, and C, have a rate of change of 2.

## OTHER TASKS

- What will count as evidence of understanding?
- What misconceptions might you find?
- What will you do or how will you respond?

 Visit this book's companion website at **resources.corwin.com/ minethegap/6-8** for complete, downloadable versions of all tasks.

**TASK 31B:** A store wants to hire a landscaper. Marcus charges $40 per hour and $300 for plants. Shay charges $50 per hour and $200 for plants.

**Which landscaper offers the best deal? Explain your reasoning for finding the best deal.**

Task 31B provides a real-world context for comparing two functions. It serves as an appropriate precursor to studying systems of linear equations. Student responses will reveal their understanding of functional behavior. Some students will note that Shay's offer has a higher rate of change, neglecting to consider how the cost of the plants will affect the total charge. Some students will select "test values" for the basis of the comparison, such as comparing the price for 5 hours of work. Other students may graph each function in order to compare. They should identify that Shay's offer is best if she works less than 10 hours, and Marcus's offer is less expensive if he is hired for more than 10 hours of work. Our students' understanding can develop at a greater rate when they exchange and compare strategies. This task would work well to support those conversations as the students described here would each benefit from different insights and approaches.

**TASK 31C:** Compare the functions $y = 3x - 1$ and $y = 4x - 3$.

**How are the functions similar?**

**Which function has a greater rate of change?**

In this task, students are provided with equations for two linear functions. Some students will be able to make comparisons based solely on this representation. Some of these students may make faulty claims about the equations themselves by using limited substitutions or by leaning on subtraction as the indicator of change. Other students may choose to graph each function before making a comparison. While students may be able to identify differences between functions, they sometimes have difficulty articulating the similarities between members of the same function family. We should help students recognize that both functions are linear and that each has a constant rate of change. Additional work with these concepts and discussion about them can help to do that. Students who make the assumption that the second function ($y = 4x - 3$) has the greater rate of change should use some sort of evidence to support their claim.

**TASK 31D:** Compare the functions $y = 2x + 5$ and $y = x^2 + 1$.

## How are the functions similar? How do the rates of change for each function compare?

Similar to task 31C, this task requires students to compare functions represented as equations. Unlike the previous task, students must now compare a linear and nonlinear function. *This task begins to expose students to other function families and requires students to begin to consider how to describe rate of change for nonlinear functions.* Some students will be able to articulate that the linear function has a constant rate of change, whereas the quadratic function has a rate of change that increases. In fact, the quadratic function's rate of change increases by a constant rate. Some students may have difficulty describing the rate of change for a nonlinear function. They may assume that there is no rate of change because the graph changes direction. We should be sure that students deeply understand the rate of change and linear functions before introducing the new type of function. Knowing that certain pieces are in place helps us determine specifically what ideas need to be developed when errors occur.

**CONTENT CONNECTION**

Early exposure to nonlinear functions helps to build understanding of rate of change as a defining characteristic of a function family. This concept helps to build a base of understanding for the study of functions in algebra and calculus.

NOTES

# BIG IDEA 32

## Systems of Equations

### TASK 32A

**Examine the scales below.**

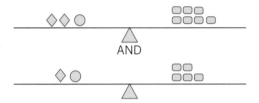

**Using the shapes provided, draw another scale that is balanced.**

**Explain how you know that this scale is balanced.**

### About the Task

There are a whole host of applications for systems of equations. But first we have to understand how systems of equations work. Balance scales provide a visual representation to support conceptual understanding of systems of equations. Students build on their previous understanding of writing and solving equations to develop an understanding of how to generate unique equations that have the same solutions. Students must analyze how the two balanced scales are interrelated in order to generate a new balanced scale. This is a necessary precursor to understanding why linear combination works as a strategy for solving systems of equations.

## PAUSE AND REFLECT

- **How does this task compare to tasks I've used?**

- **What might my students do in this task?**

 Visit this book's companion website at
**resources.corwin.com/minethegap/6-8**
for complete, downloadable versions of all tasks.

## Anticipating Student Responses

Our students may use a variety of techniques to generate a new equation. Students may apply the technique of multiplying by a scalar to generate a new balanced scale. One example of this technique may be two diamonds and two circles balance with ten rectangles. Other students may compare the differences between the two scales to determine how they have changed and then apply that change to their new equation. These students may add the same quantity to each side of the scale, such as representing one diamond and two circles balancing with five rectangles and one circle. Students may also add or subtract the sides of the balance to produce a new balanced scale. Students may show that three diamonds and two circles are balanced with twelve rectangles.

NOTES

### Student 1

Student 1 decides to assign values for the weight of each shape at random. He assigns the diamond to have a value of 10, the circle to have a value of 5 and the rectangle to have a value of 5. However, Student 1 does not confirm his results correctly for each of the scales. Student 1 doesn't explain how he knows that the scale is balanced.

### Student 2

Student 2 notices that the difference between the two scales is the number of diamonds and rectangles. He uses this information to conclude that the diamond must have the same weight as two rectangles. He shows this on his created scale. Student 2 provides a clear explanation for his thinking.

**USING EVIDENCE**

*What would we want to ask these students? What might we do next?*

### Student 1

Student 1's strategy can be useful, but he needs to focus on the relationship between the shapes. We should ask how the first two balance scales are similar and how they are different. We might ask if the scale would remain balanced if all of the shapes were combined. We might draw his attention to the similarities and differences between the scales. He should see that the top scale has one more diamond and two more rectangles. We can thus ask what two diamonds are equivalent to and see if he is able to use this information to determine the value (in rectangles) of the circle. We might then have him substitute values to determine if his conclusions are accurate.

### Student 2

First, it is important to acknowledge that Student 2 has applied an appropriate strategy for generating a new scale. He correctly noticed that by finding the difference of both sides, he would find a resulting scale that would be balanced. Student 2 now needs the opportunity to expand his thinking to other possible operations with the scales. We can summarize his approach and restrict the possibilities for the new creation. For example, we might say the new scale has more than two diamonds or more than two rectangles.

## Student Work 1

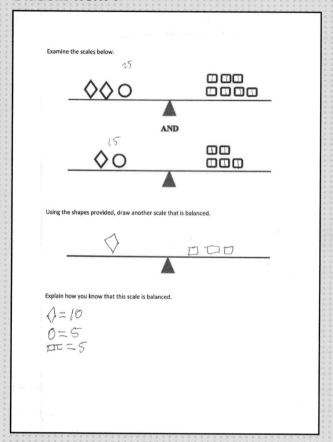

## Student Work 2

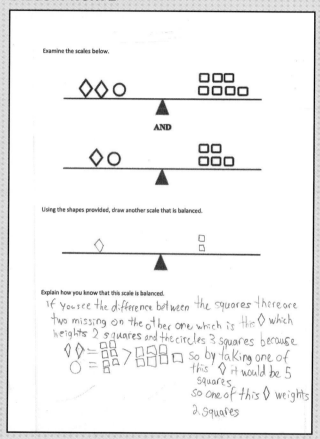

### Student 3

Student 3 added the two balances together to produce a new balanced scale. He found that three diamonds and three circles are equivalent to twelve rectangles. Student 3, however, does not describe that he found the sum of the scales. He only states that a diamond is equal to two rectangles and a circle is equal to three rectangles, but does not provide an explanation of how he made this conclusion.

### Student 4

Student 4 has also produced a scale that is balanced. His work is similar to that of Student 2. The strategy used to make this scale is not clear from the student's explanation. He states, "I know that a diamond is 2 squares so a circle is 3 squares." He provides no explanation as to how he determined the relationship between the three shapes.

***What would we want to ask these students? What might we do next?***

### Student 3

Like Student 2, Student 3 has applied an appropriate strategy for generating a new scale. He correctly noticed that by finding the sum of both sides, he would find a resulting scale that would be balanced. However, Student 3 has not explained how he determined some of his conclusions. We should ask him to explain how he generated the new scale and how he determined the relationship between the shapes. Noting that he found the sum of the two scales, we can ask him to apply what he knows about the relationship to produce other scales. We should look to see if his new creations continue to add known values to both sides or if he is able to create a scale with an isolated shape on the left.

### Student 4

Student 4 has also produced a balanced scale. Like Student 3, Student 4 has not fully explained his conclusions. We should redirect him to explain how he generated the new scale and how he determined the relationship between the rectangles and the other shapes. Next steps for Student 4 are likely similar to those for Students 2 and 3. It may also be useful to have subsequent tasks use real-world items as a context for balancing scales. For example, we might replace the shapes with different types of fruit or something similar. Doing so supports how systems of equations apply to real-world problems.

## Student Work 3

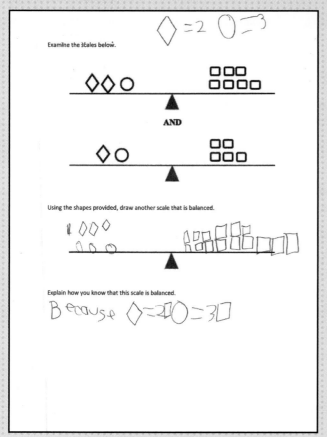

## Student Work 4

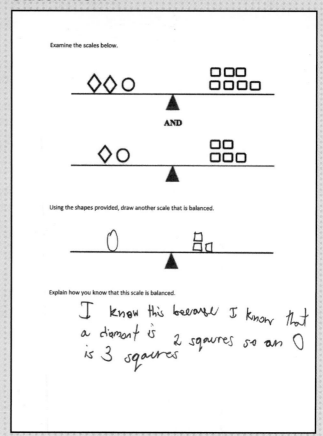

## OTHER TASKS

- What will count as evidence of understanding?

- What misconceptions might you find?

- What will you do or how will you respond?

 Visit this book's companion website at **resources.corwin.com/ minethegap/6-8** for complete, downloadable versions of all tasks.

**TASK 32B:** Aurora is making jewelry to sell at the craft fair.

- On Monday, Aurora was able to make 3 necklaces and 5 bracelets in 5 hours 20 minutes.

- On Tuesday, Aurora was able to make 1 necklace and 3 bracelets in 3 hours.

**On Wednesday, Aurora needs to make 2 necklaces and 1 bracelet. How much time will it take her to finish?**

This task requires students to apply understanding of systems of equations to a real-world context. This scenario does not directly require that students find how long it takes to produce each item. Rather, it requires students to determine the time it takes to make 2 necklaces and 1 bracelet. Students may write and solve a system of equations for Monday and Tuesday, then substitute the solution into an equation for Wednesday to find the solution. A common student error is to find the time it takes to make each item, but not respond to the question posed in the task.

**TASK 32C:** The local movie theater offers the following snack packages:

$13.50

$20.00

*Image source:* Popcorn: © Clipart.com; Soda: © Pixabay.com/OpenClipart-Vectors

### Find the cost of each individual item.

*Similar to task 32A, this task provides visual representation of a system of equations.* It may be used to develop conceptual understanding, especially because it is in a somewhat familiar context. This task may be given prior to students learning

**MINING TIP**

Representations help develop conceptual understanding of mathematics ideas. Connecting representations and contexts may be most effective. When doing so, we should be sure to connect equations even if our students aren't ready to work with them directly.

how to use algebraic strategies for solving equations. Students may use a variety of strategies including linear combination and substitution in order to solve. For example, students may recognize that the second snack has one more popcorn than the first snack. The difference in cost of the two plans is $6.50, which would represent the cost of this extra popcorn. Students may then use substitution to conclude that each soft drink costs $3.50.

**TASK 32D:** Solve the following system of equations.

$$3x + 6y = 12$$
$$x + 2y = 4$$

**Next, write another equation that has the same solution as this system.**

*While this task seems like a traditional system of equations, the task also requires students to write a new equation that will yield the same solution.* Students may use a variety of strategies to solve, including linear combination, substitution, or graphing. Students using linear combination may get confused when they produce the equation $0 = 0$. Students may fail to realize that these two equations are representations of the same line, since the first equation is the second equation multiplied by a constant of 3. Some students may reason that one equation has been changed by a scale factor of 3. These students might then multiply either equation by a different scalar to produce a new equivalent equation.

**MINING TIP**

It is important to take note of the representations that our students use. Predominant use of the same representation or strategy may be indicative of limited understanding or biased approaches during instruction.

NOTES

# CHAPTER 6

# GEOMETRY

## THIS CHAPTER HIGHLIGHTS HIGH-QUALITY TASKS FOR THE FOLLOWING:

- Big Idea 33: Area of Composite Figures

  We can find the area of figures that are more complex than regular polygons by decomposing or composing these figures to find composite area.

- Big Idea 34: Nets and Three-Dimensional Figures

  Three-dimensional shapes can be represented with two-dimensional diagrams. Understanding nets and their relationship to three-dimensional figures helps us understand volume, surface area, and planar cross sections.

- Big Idea 35: Surface Area and Volume

  Surface area and volume have countless real-world applications. The dimensions of a three-dimensional object can be found to maximize or minimize the surface area and volume of that object.

- Big Idea 36: Volume of Cylinders and Cones

  Volume of cylinders and cones also has real-world application. Finding the volume of these figures is different. Altering their dimensions will result in an increase or a decrease in the volume of that object.

- Big Idea 37: Angle Relationships

  Angle relationships are connected through lines and intersections. We can reason about these relationships to determine unknown measures and other quantities.

- Big Idea 38: Transformations, Similarity, and Congruence

  Transformations describe how we can manipulate shapes. All transformations produce similar figures in the plane. Some transformations will produce congruent figures.

- Big Idea 39: Distance and the Pythagorean Theorem

  The Pythagorean Theorem provides a means for finding the lengths of segments, including the distance between any two points.

# BIG IDEA 33
## Area of Composite Figures

### TASK 33A

> In the figure, the length of segment AB is 4 inches.
>
> Find the area of the shaded region. Explain your reasoning.

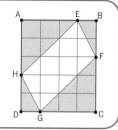

**CONTENT CONNECTION**

Area is revisited in various contexts in the study of algebra, geometry, and calculus. Students apply similar reasoning to solve problems including estimating the area below curves and calculating the accumulation of functions.

### About the Task

Area is all around us. It has many practical applications. It is an important concept in mathematics as it connects to other big ideas. As we know, there is much more to it than finding the area of rectangles or squares. *Finding the area of composite figures provides students with the opportunity to apply reasoning and critical thinking skills.* We can mix our understanding of area and reasoning to solve complex problems. In this task, students have the opportunity to make connections between the area of two figures. They will also apply reasoning to find the area of a shaded region by combining partial shading of different square units.

### Anticipating Student Responses

Our students are likely to approach this problem in a few different ways. Some students will see that the shaded region consists of four triangles. These students will find the sum of each triangle's area. Other students may see the shaded region as the difference between the large rectangle and the nonshaded parallelogram. Students who use this strategy may mistake the side length of the parallelogram with the height of the parallelogram. Other students may use a variety of estimation strategies by considering squares that are completely shaded and estimating those that are partially shaded. It is possible that students may cut the picture and rotate regions to form complete boxes.

### PAUSE AND REFLECT

- How does this task compare to tasks I've used?
- What might my students do in this task?

 Visit this book's companion website at **resources.corwin.com/minethegap/6-8** for complete, downloadable versions of all tasks.

### Student 1

Student 1 finds the area of rectangle ABCD to be 20. She incorrectly reasons that the shaded region is half of the total area, so she divides the total area in half to find a solution of 10. *Student 1 doesn't label her solution in the appropriate units and is unable to explain her reasoning.*

### Student 2

Student 2 applies an estimation strategy to find the area of the shaded region. She marks squares that are completely shaded and found that there are six. Student 2 finds six squares that are half-shaded. She pairs the two small triangles, noting the sum of the pairings would make two additional squares. Student 2 then finds the total area of the shaded region to be 11 square inches, though she incorrectly writes "112 in."

**USING EVIDENCE**

*What would we want to ask these students? What might we do next?*

### Student 1

We must determine if and why Student 1 believes the parallelogram to be half of the shaded region. She may believe it because she perceives two shapes in the figure or possibly because she confused division with subtraction from the whole. We could approach the error by having her cut out the figures and compare their area. Before focusing on area explicitly, we want to be sure that she can compose and decompose shapes. Assuming she can, we might ask her to decompose the shaded figure into composite shapes and discuss their attributes and dimensions. Alternatively, we could focus the discussion on the relationship between the parallelogram and rectangular shape of the shaded region. We may also consider additional work with less complicated composites before returning to work with this and similar tasks.

### Student 2

Student 2 applies a valid estimation strategy to find the area of the shaded region. She can combine regions to form areas that she knows. She has a novel method for noting related shapes. We do want to help her communicate her reasoning clearly, particularly with explaining how she determined the half-square areas and the regions she marked with a circle or small square. We should leverage her work during whole class debriefing so that others may develop their ideas about irregular areas. We can advance her understanding by providing more complex representations for her to consider.

**TASK 33A:** In the figure, the length of segment AB is 4 inches.

Find the area of the shaded region. Explain your reasoning.

## Student Work 1

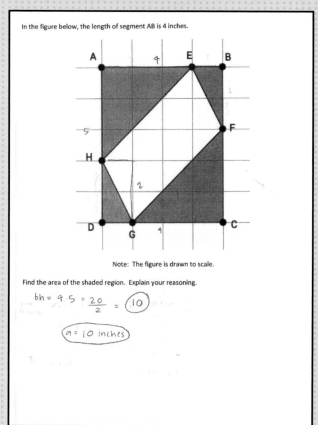

In the figure below, the length of segment AB is 4 inches.

Note: The figure is drawn to scale.

Find the area of the shaded region. Explain your reasoning.

$bh = 4.5 = \dfrac{20}{2} = \boxed{10}$

$a = 10$ inches

## Student Work 2

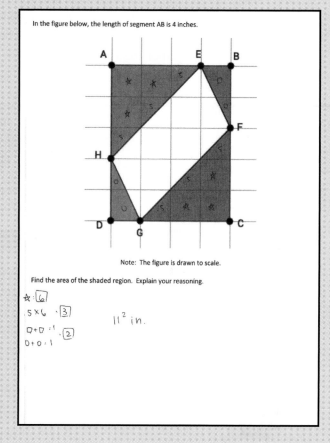

In the figure below, the length of segment AB is 4 inches.

Note: The figure is drawn to scale.

Find the area of the shaded region. Explain your reasoning.

☆ : $\boxed{6}$

$.5 \times 6 \quad . \boxed{3}$

$\square + \square : ^1 . \boxed{2}$

$0 + 0 : 1$

$11^2$ in.

### Student 3

Student 3's strategy finds the area of the parallelogram and subtracts it from the area of the larger rectangle. While her strategy is valid, Student 3 incorrectly calculates the area of parallelogram EFGH. This happens because she finds faulty lengths of GH and EF. She may find this length as the result of half of the sum of the lengths of GD and DH. She determines that the area of EFGH is the product of the side lengths, rather than the product of the base and height.

### Student 4

Like Student 3, Student 4 applies the strategy of subtracting the area of EFGH from the area of ABCD. However, Student 4 decomposes the parallelogram into five triangular regions. This enables Student 4 to find the sum of these areas and subtract it from the area of rectangle ABCD.

**USING EVIDENCE**

*What would we want to ask these students? What might we do next?*

### Student 3

**MINING TIP**

**Some students working with this task may not yet have been introduced to the Pythagorean Theorem. We can modify the task to show the height of the parallelogram to support their work.**

Student 3 has a valid approach to solving the problem, which we should acknowledge. However, she has flawed understanding relative to the attributes and area of a parallelogram. We should first work to confirm that she understands how to find the area of triangles and quadrilaterals. *We also need to explore why it is difficult to find the height of the parallelogram in this task.* We might challenge her to decompose the diagram into various triangles and squares or ask if she recognizes the shapes of the shaded area. If so, we could have her find the difference of the area of the large rectangle and the sum of the shaded triangles.

### Student 4

Student 4 finds the area of the parallelogram without necessarily knowing the formula for the area of a parallelogram or the height of the shape. Student 4 and Student 3 should compare their approaches to see how they are similar and different. But to do so, she will need to elaborate on her thinking in order to communicate her reasoning to her peers. We might connect her ideas of composite triangles with the apparent shaded triangles in the figure. We can ask her how she might use those shaded triangles to find a solution to the task. When doing so, it's important to acknowledge her approach so that we don't insinuate that there is a better approach.

**TASK 33A:** In the figure, the length of segment AB is 4 inches.

Find the area of the shaded region. Explain your reasoning.

## Student Work 3

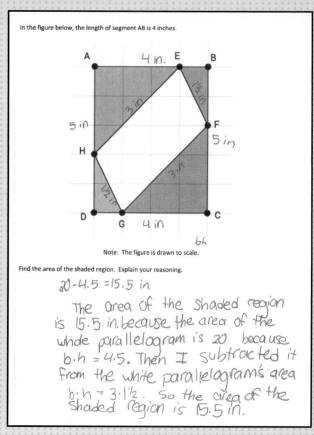

In the figure below, the length of segment AB is 4 inches.

Note: The figure is drawn to scale.

Find the area of the shaded region. Explain your reasoning.

20 − 4.5 = 15.5 in.

The area of the shaded region is 15.5 in. because the area of the whole parallelogram is 20 because b·h = 4.5. Then I subtracted it from the white parallelogram's area b·h = 3·1½. So the area of the shaded region is 15.5 in.

## Student Work 4

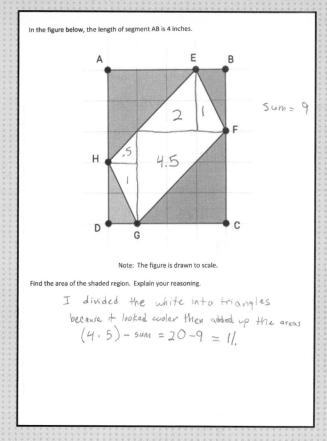

In the figure below, the length of segment AB is 4 inches.

Sum = 9

Note: The figure is drawn to scale.

Find the area of the shaded region. Explain your reasoning.

I divided the white into triangles because it looked cooler then added up the areas

(4·5) − sum = 20 − 9 = 11.

## OTHER TASKS

- What will count as evidence of understanding?
- What misconceptions might you find?
- What will you do or how will you respond?

Visit this book's companion website at **resources.corwin.com/minethegap/6-8** for complete, downloadable versions of all tasks.

**TASK 33B:** In the figure, the length of segment AF is 3 cm.

**Is the area greater or less than 45 cm²? Explain your reasoning.**

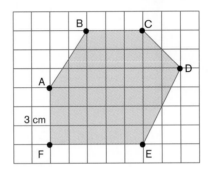

In this task, students will need to estimate the area of the polygon. It is important to note that the task does not require students to find the area. They simply must determine if the area is greater or less than 45 square centimeters. Some students may decompose the figure into rectangles and triangles and find the sum of the areas of the composite shapes. Others might notice that the polygon is within a rectangle and they will choose to find the area of the rectangle and compare it to the embedded polygon. For these students, we should be sure to ask how they answered the question without finding the area.

**TASK 33C:** Create a composite figure composed of triangles and quadrilaterals on the grid.

**Find the area of your figure. Explain how you determined your answer.**

*This open-ended task allows students to apply their understanding of composite figures in order to find the area of their own figure.* Students may create a closed figure by adjoining the shapes. Some students may choose to use shaded regions for some of the figures to distinguish between the figures. This task also provides opportunities to spot and address misconceptions. Students may incorrectly find the height of non-right triangles. Students may choose to draw a trapezoid or parallelogram. Common errors with these shapes include using side length rather than height to find the area.

**MODIFYING THE TASK**

We can modify the task to ask students to create a composite figure with a specific, targeted area. We can then compare student creations and discuss how different figures can have the same area.

**TASK 33D:** The figure below displays two circles inscribed in a rectangle. The length of segment IL is 8 cm, and the length of segment MO is 4 cm. Find the area of the shaded region.

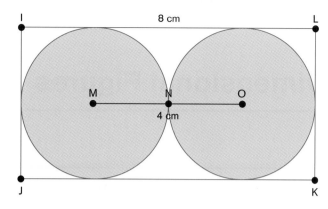

**Note: The figure is drawn to scale.**

*Task 33D requires students to have some understanding of the area of a circle.* Most students will find the area of the circles M and O and then subtract from the area of rectangle IJKL. Common student errors will include having a circle radius of 4 cm, which is the diameter. They may not realize the lengths of IJ and KL are equivalent to the diameter of the circle and only subtract the area of one circle from the rectangle. Students may also incorrectly identify the units of measure as centimeters rather than square centimeters.

**CONTENT CONNECTION**

The area of circles is typically introduced in Grade 7. This task can serve as a connection between circles and composite figures once students have an understanding for finding area of circles.

NOTES

# BIG IDEA 34
## Nets and Three-Dimensional Figures

### TASK 34A

**Tell how you know if each represents a prism.**

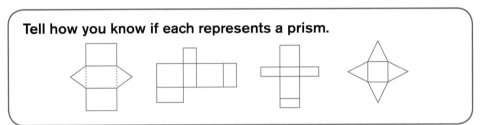

### About the Task

Nets are two-dimensional representations of three-dimensional figures. Understanding nets and their relationship to three-dimensional figures connects and reinforces our students' understanding of volume, surface area, and planar cross sections. These understandings are critical for future work in geometry and calculus. Nets can help students reason about three-dimensional figures and even decompose them. In this task, students consider and explain if each net will produce a prism. Typically, students are asked to identify what figure a specific net creates. In this task, students compare and contrast different nets, considering inclusive attributes of a prism.

### Anticipating Student Responses

Students will analyze the nets in different ways. One common strategy is to focus on the types of shapes in the net, such as number of rectangles, number

### PAUSE AND REFLECT

- How does this task compare to tasks I've used?
- What might my students do in this task?

Visit this book's companion website at
**resources.corwin.com/minethegap/6-8**
for complete, downloadable versions of all tasks.

of triangles, and relative size of each shape. Students may mistakenly believe that all prisms are rectangular prisms, and so they will eliminate the two nets that contain triangles. Using this same reasoning, students may mistakenly identify the upper right net as a prism because it consists of rectangles, overlooking the relative size and shape of each face of the net. *Some students may attempt to try to draw or construct the three-dimensional figure for each given net.* Students may be challenged to draw a two-dimensional picture that represents a three-dimensional object.

**MINING TIP**

**It may be helpful to have clay or something similar for students to manipulate into three-dimensional objects as students begin work with these concepts.**

NOTES

### Student 1

Student 1 uses his definition of a "closed figure" as the basis for determining whether each net will produce a prism. He determines that the second net is the only net that will not produce a prism, "because it isn't a closed figure." It appears that Student 1 is defining "closed figure" by whether the net will fully cover a three-dimensional object without any holes or gaps. The second net has one small square side, which would leave a gap on one side.

### Student 2

Student 2 applies different criteria for each of the nets. He correctly identifies the first net as a prism. Apparently, he visualizes what the net would cover. Student 2 then correctly states that the second net is not a prism, but reasons that "its more flat on top" without any clarification. He disagrees with the third as it is a cube while the last is a prism because "it forms together at the top."

## USING EVIDENCE

*What would we want to ask these students? What might we do next?*

### Student 1

First, we need Student 1 to clarify what he means by "closed figure." We should reinforce that nets for any three-dimensional figure must "close." We can then ask him to compare and contrast different examples of prisms and pyramids. We may even add cylinders and cones. We can leverage his ideas about their similarities and differences to establish what a prism is. When doing so, we should reinforce that each is "closed" as well as their differences. With understanding of these different three-dimensional figures in place, we can move to nets.

### Student 2

Like Student 1, Student 2 also seems to notice that the second net will not create a three-dimensional figure. We should also clarify the criteria used for determining prisms. We need to understand what he means by "flat on top" and forming at the top (last net) and how this thinking determines if a net is a prism. His reference to a cube for the third net signals some understanding of types of three-dimensional figures but also indicates misunderstanding about the relationship between cubes and prisms. Like Student 1, Student 2 seems to have unfinished learning about types and attributes of different three-dimensional figures.

## Student Work 1

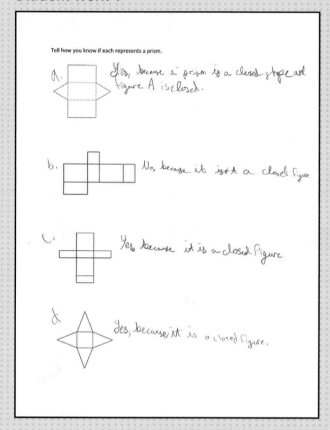

Tell how you know if each represents a prism.

a. Yes, because a prism is a closed shape and figure A is closed.

b. No, because it isn't a closed figure.

c. Yes because it is a closed figure.

d. Yes, because it is a closed figure.

## Student Work 2

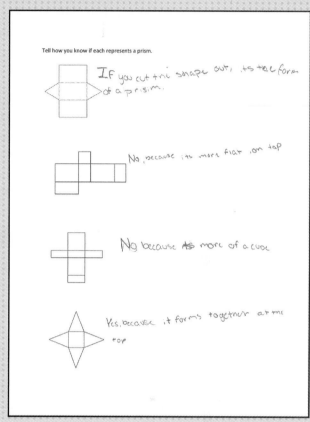

Tell how you know if each represents a prism.

If you cut the shape out, its the form of a prism.

No, because its more flat on top

No because its more of a cube

Yes, because it forms together at the top

### Student 3

Student 3 uses different reasons for determining whether a net produces a prism. He determines that the first net will not produce a prism "because it has triangles in it." However, this reasoning does not extend to the fourth net, which he does consider to be a prism. Student 3 also considers if nets will create "closed shapes" and applies that reasoning unevenly to the second and third nets.

### Student 4

Student 4 assesses each net based on the number of rectangles and triangles it has. He concludes that the net produces a prism if it "is mainly made out of rectangles." Student 4 concludes that the net will not produce a prism if it "is mainly made out of triangles." While this logic works for most nets in this task, Student 4 overlooks the relative size of each rectangle, particularly for the second net.

*What would we want to ask these students? What might we do next?*

### Student 3

Student 3 offers contradictions that we must explore. He accurately considers the openness of a figure but is unable to apply it correctly to both the second and third nets. He notes that the first net is not a prism because of its triangles but then agrees that the last net (with triangles) will create a prism. Interestingly, he does include an idea about the evenness of the sides of the fourth net. Like the others, he shows that he has unfinished learning about the attributes of prisms and three-dimensional figures in general. He needs more opportunities to consider the attributes of various figures. We should also be sure to spend time folding nets and comparing them to known figures. We can use paper models or commercial products that feature plastic nets that can be folded and placed into clear plastic figures.

### Student 4

We can build on Student 4's belief that prisms are built with rectangles. But we want him to understand why his strategy of considering rectangles and triangles doesn't always work. We could have him sort prisms and pyramids into (only) two different groups. Essentially, we have to establish that prisms can also have a limited number of triangular faces. We might also ask him how he knows that having more triangles will not produce a prism. Student 4 will need to articulate that prisms are elongations of one face. Because of this, there can be no more than two triangular faces for the net of a prism. He and the other three students are good examples of student work for class discussion. Each has unique ideas that, when related, may establish a better collective understanding of the concept of nets and prisms.

## Student Work 3

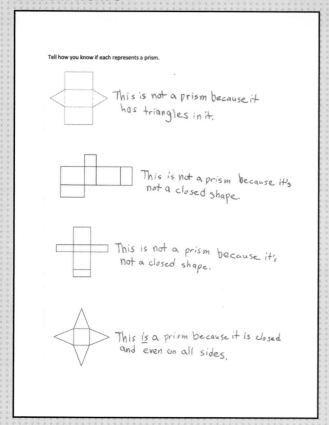

Tell how you know if each represents a prism.

This is not a prism because it has triangles in it.

This is not a prism because it's not a closed shape.

This is not a prism because it's not a closed shape.

This is a prism because it is closed and even on all sides.

## Student Work 4

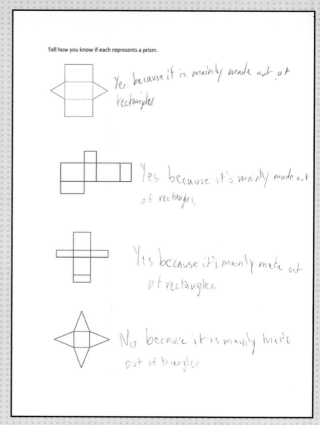

Tell how you know if each represents a prism.

Yes, because it is mainly made out of rectangles

Yes because it's mainly made out of rectangles.

Yes because it's mainly made out of rectangles.

No because it is mainly made out of triangles

## OTHER TASKS

- What will count as evidence of understanding?
- What misconceptions might you find?
- What will you do or how will you respond?

Visit this book's companion website at **resources.corwin.com/minethegap/6-8** for complete, downloadable versions of all tasks.

**TASK 34B:** Derek noticed these two nets.

**How are the two nets similar and how are they different?**

**Tell why both nets will or will not make the same three-dimensional object.**

Task 34B requires students to analyze and compare nets for two types of pyramids. They should note that both nets will produce a pyramid, though one has a square base and the other has a triangular base. They should relate that the pyramids have a different number of triangular faces. Some students may believe that one of the figures produces a prism because it has a square base. Others may believe that it cannot be a pyramid because it has a square or because the other is made solely of triangles.

**TASK 34C:** Examine the net. Describe the three-dimensional object that the net represents. Is it possible to construct a different net for this three-dimensional object? Why or why not?

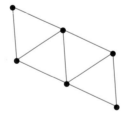

Students can be led to misconceptions about mathematics concepts when we overly rely on specific representations of those concepts. Often, students are provided a net with one shape in the center and spokes emerging from this center shape. This task features an unconventional representation of a net for a triangular pyramid. This alone may lead students to discount the validity of this net. Those who do overlook it may indicate that they look for a pattern or familiar diagram rather than consider how the net may be folded. Successful students will realize that this net will produce a triangular pyramid.

**TASK 34D: Construct two different nets that would produce the same rectangular prism.**

Our students can develop the misconception that a figure can only be created from a specific net. This task challenges them to consider how the same prism could be formed by two different nets. Some may believe that it simply isn't possible. Others may create two different nets that do not create the same prism. Students will need to realize that similar rectangles will need to be used to produce the faces of the prism, but the location of rectangles may be relocated. Options may be limited due to the locations of the faces. The case that produces the most options will be a square top and bottom face, because students will be able to adjoin these faces to the top and bottom of any of the rectangular sides.

NOTES

# BIG IDEA 35
## Surface Area and Volume

### TASK 35A

Create a diagram for three different rectangular prisms that each have a volume of 64 cm³.

How do their surface areas compare?

**CONTENT CONNECTION**

Area and volume are explored in algebra and calculus through the context of maximizing volume and minimizing surface area or cost. This task provides an opportunity to touch on these ideas by simply comparing results and commenting on the appearance of the prisms.

### About the Task

Volume is an important concept in mathematics. Misconceptions between volume and surface area are similar to area and perimeter in that students think the two are directly related to one another. Task 35A begins to help students differentiate between volume and surface area. Students must create three different representations of rectangular prisms that have a volume of 64 cubic centimeters. Then, they must compare the surface areas of these figures. Students should notice manipulation of the side lengths affects the surface area. *The prism that is closest to being a cube will have the smallest surface area.*

### Anticipating Student Responses

Most students will begin this task by finding a set of dimensions that will produce a volume of 64 cubic centimeters. Some students may begin halving and doubling the dimensions to find other sets of dimensions. Some students will create diagrams that are approximately to scale, showing elongated or compacted rectangular prisms based on the dimensions. Misconceptions

### PAUSE AND REFLECT

- How does this task compare to tasks I've used?
- What might my students do in this task?

 Visit this book's companion website at **resources.corwin.com/minethegap/6-8** for complete, downloadable versions of all tasks.

may emerge as students analyze the surface area. Some students may incorrectly assume that the surface area and volume are equal. ***Students may simply calculate the area of one side for comparison rather than comparing surface areas.*** A variety of calculation errors may occur. Some students may draw the net for each prism to compare the surface areas rather than calculate.

**MINING HAZARD**

**This error provides an opportunity to revisit the concept of nets as a strategy for calculating surface area.**

NOTES

### Student 1

Student 1 finds the cube root of 64, but it does not appear that she applies this dimension to any of her rectangular prisms. Instead, she uses dimensions like 2 cm × 3 cm × 1.5 cm for the rectangular prism, even though these dimensions will not produce a volume of 64 cubic centimeters. Student 1 makes slight variations to these dimensions to produce other incorrect solutions. She concludes that the surface areas "all have nearly the same dimensions."

### Student 2

Student 2 creates three rectangular prisms with different dimensions that have a volume of 64 cubic centimeters. Her diagrams are not to scale. For example, the third rectangle should be the most elongated prism. She shows the classic misconception with her statement "Because they all have the same volume, their surface area is the same."

*What would we want to ask these students? What might we do next?*

### Student 1

We can use Student 1's ability to draw prisms and find a cube root to better understand how she determined the dimensions of her prisms. We should ask her to find the volume of her prisms and compare her findings to the required 64 cubic centimeters. It may be that her ideas about prisms and finding volume are disconnected. It may be that she sees the mathematics as pure procedure without meaning. We can leverage her ability to find a cube root computationally and ask her to compare that to the dimensions of a related prism. We may also consider having her simply create prisms with different volumes and surface areas before providing a condition of 64 cm$^3$ for three different prisms. Doing this will help us further investigate her understanding of these concepts.

### Student 2

Student 2 can create prisms for a given volume. *We must now work to disconnect the perceived relationship between volume and surface area.* We can provide her with nets of different prisms that have the same surface area or volume. We can give her the opportunity to find the area and volume of these nets. We may even have to let her cut out and build the prisms to reinforce her conclusions. It's likely that this investigation will need to be repeated so that she can fully develop this understanding. It is wise to bundle the investigations with class discussion of results, strategies, and observations.

**MINING TIP**

We can be tempted to tell students that the two aren't related and then provide examples or charts to verify our argument. However, for long-term understanding, our students have to discover these ideas and make the conclusions themselves.

## Student Work 1

Volume = Length x Width x Height

$\sqrt[3]{64} = 4 cm$

Create a diagram for three different rectangular prisms that each have a volume of 64 cm³.

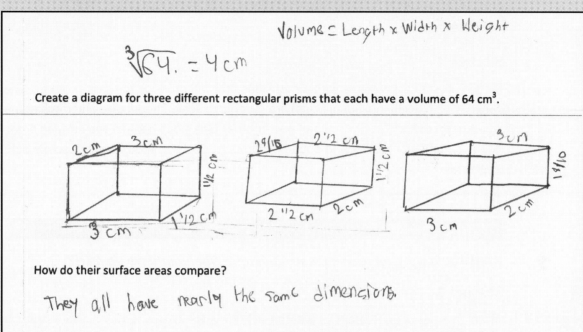

2 cm    3 cm    1½ cm    3 cm

29/15   2½ cm   1½ cm   3 cm   19/10

3 cm    1½ cm   2½ cm   2 cm   3 cm   2 cm

How do their surface areas compare?

They all have nearly the same dimensions.

## Student Work 2

Create a diagram for three different rectangular prisms that each have a volume of 64 cm³.

4 cm   8 cm   2 cm   $V = 64 cm^3$

8 cm   8 cm   1 cm   $V = 64 cm^3$

2 cm   2 cm   16 cm   $V = 64 cm^3$

How do their surface areas compare?

Because they all have the same Volume, their surface area is the same

## Student 3

Student 3 creates prisms with a volume of 64 cm³. We should note that her first two prisms are the same with different orientations. Student 3 omits units of measure. Student 3 only finds the area of one of the sides of the prism rather than calculating the surface area.

## Student 4

Student 4 generates prisms with dimensions that will produce the same volume. The dimensions of each individual prism are roughly to scale, though the diagrams of the different prisms are not to scale of one another. She finds each of the surface areas for the rectangular prisms and applies appropriate units of measure to describe side lengths and area.

**USING EVIDENCE**

*What would we want to ask these students? What might we do next?*

## Student 3

When working with Student 3, we can acknowledge that she has a valid strategy for finding dimensions of rectangular prisms with the same volume. We should ask her how her first two prisms are similar and different. We want her to establish that they are, in fact, the same. A physical model of the prism(s) may be helpful. We can then charge her with creating a new, different prism that satisfies the prompt. Our heavy lifting with Student 3 is to help her understand what surface area is. She may confuse the idea with area. We can take advantage of her ability to find area of two-dimensional shapes. We can have her work with nets of these prisms to develop understanding of surface area.

## Student 4

Student 4 demonstrates understanding of volume and surface area. The focus of work for her will be to refine her comparison and to extend her thinking. We can ask her if the diagrams provide an accurate picture for comparing the three prisms. We can highlight their relative size to one another. We want her to recognize that the second rectangular prism should be smaller in size. We can provide other opportunities to help her develop a sense of scale. We should also encourage her to reference the difference between volume and surface area in some way. We might extend her thinking by asking, "What do you notice about the relationship between the size of the diagram and the surface area?" or "How do the dimensions of the prism impact the surface area?"

## Student Work 3

Create a diagram for three different rectangular prisms that each have a volume of 64 cm³.

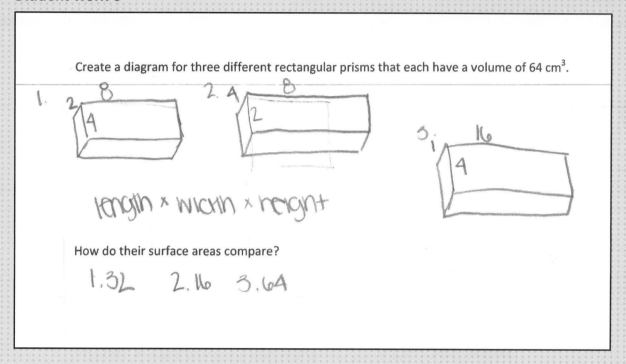

length × width × height

How do their surface areas compare?

1. 32    2. 16    3. 64

## Student Work 4

Create a diagram for three different rectangular prisms that each have a volume of 64 cm³.

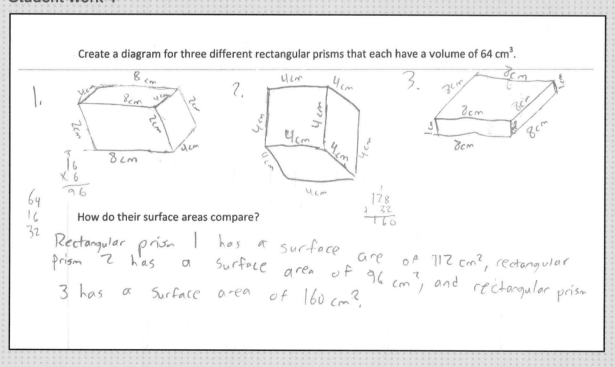

How do their surface areas compare?

Rectangular prism 1 has a surface area of 112 cm², rectangular prism 2 has a surface area of 96 cm², and rectangular prism 3 has a surface area of 160 cm².

OTHER TASKS

- **What will count as evidence of understanding?**
- **What misconceptions might you find?**
- **What will you do or how will you respond?**

 Visit this book's companion website at **resources.corwin.com/ minethegap/6-8** for complete, downloadable versions of all tasks.

**CONTENT CONNECTION**

**This extension provides another opportunity to preview the concept of optimization through comparison of models. Students will most likely not find the exact dimensions to maximize volume, but should be able to find dimensions that produce larger volumes.**

**TASK 35B:** **You have a 24 in. × 18 in. sheet of wrapping paper to wrap a birthday gift. What size box would you be able to wrap with this sheet of wrapping paper? Justify your reasoning.**

This task complements task 35A. In it, students begin with a sheet of wrapping paper and must consider how to apply nets of rectangular prisms to find surface area and volume. Students may choose to draw a net within a rectangle with the dimensions of the wrapping paper. Students may apply trial and error to find dimensions that work. Some students may make the error of using the area of 432 as a volume to find the dimensions, rather than considering the surface area. *As an extension, we might consider asking students what would be the dimensions of the box with the largest volume that could be wrapped with this wrapping paper.* This extension allows students to compare different rectangular prisms in order to compare volumes.

**TASK 35C:** **The triangular prism has a surface area of 180 cm².**

**What might be the dimensions of the prism? Explain your reasoning.**

Task 35C is an opportunity to apply understanding of surface area to a triangular prism. Students will need to consider that a triangular prism has sides that are rectangles and triangles. Students may draw a net to assist with their work. Some students may inaccurately think of the surface area as the volume and try to divide 180 into different factors. Successful students might realize that they need to find dimensions for a rectangle and triangle that, when tripled and doubled respectively, create a surface area of 180 square centimeters. Some students may consider a right triangular prism rather than the triangular prism shown. We may also find successful students who simply "crunch numbers" without regard to the relative size or lengths represented in the diagram. We should be sure to look for this as it may not reveal understanding of surface area, though correct values may be found.

**TASK 35D:** The shed pictured here needs the four exterior sides to be painted.

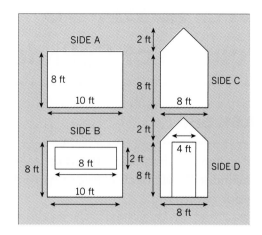

*Image source:* House: © iStockphoto.com/MarkUK97

**If a gallon of paint covers 200 square feet of a surface, how many gallons of paint are needed to paint the sides of this shed? Be sure to justify your solution with pictures, numbers, or words.**

Most students will work within the diagrams to calculate the surface area. Some students may forget to subtract the area of the windows and doors. Some may argue that the door should also be painted. Students who find the surface area may think that they are finished. We might redirect these students to the prompt. An alternative is to bring all students together to compare their solutions. As we know, some will have a solution relative to area and others will have a solution relative to gallons of paint. We can position them to make choices about what the solution should be due to the conversations that they have. Those who find gallons may provide a decimal answer. This too is an opportunity for discussion. We might extend the problem by providing costs for gallons of paint or different volumes of paint cans.

NoTES

# BIG IDEA 36
## Volume of Cylinders and Cones

### TASK 36A

> If cylinder A has twice the diameter and half the height of cylinder B, will each cylinder have the same volume? Explain your reasoning.

### About the Task

**CONTENT CONNECTION**

**This task requires students to explore how the dimensions of a three-dimensional object affect the volume. This may serve as an opportunity to explore the concept of maximizing volume, which is studied in greater depth in advanced algebra and calculus courses.**

We make assumptions about dimensions, area, and volume that relate to ideas of operations and whole number relationships. Yet, these assumptions may not always apply to concepts as we believe they should. As students build fluency with volume, they will need to extend understanding of volume to develop strategies for determining and comparing the volumes of cylinders and cones. *In this task, students must analyze the relationship between height and diameter for a cylinder to examine how modifying these values will impact the volume of the cylinder.* Students should discover that altering dimensions of a three-dimensional object will result in an increase or a decrease in the volume of that object and that the change may not be directly related to the change in one or more dimensions.

### Anticipating Student Responses

Most students will approach this task by using sample values for the diameter and height. For example, a student may describe cylinder A with a 10 cm diameter and a 10 cm height. Their cylinder B would have a diameter of 5 cm and a height of 20 cm. In this case, the student would find that the volume

**PAUSE AND REFLECT**

- How does this task compare to tasks I've used?
- What might my students do in this task?

Visit this book's companion website at **resources.corwin.com/minethegap/6-8** for complete, downloadable versions of all tasks.

of cylinder A (785.4 cm³) would be greater than that of cylinder B (392.7 cm³). A common mistake students may make is to forget to find the radius for each cylinder. Students may also incorrectly conclude that halving and doubling (as with whole numbers) will result in the same volume. This flawed conclusion occurs because they do not account for the radius to be squared.

NOTES

## Student 1

Student 1 attempts to reason about how each of the dimensions is being altered. He states, "...even if you make the cylinder smaller, you are also making it wider so the volume would be the same." Student 1 believes because the dimensions are being halved and doubled, the volume will remain the same but doesn't account for the squaring of the radius.

## Student 2

Student 2 incorrectly concludes that the volumes of both cylinders would be the same. He creates dimensions for cylinder A and then doubles and halves the dimensions for cylinder B. He uses the formula for the volume of a rectangular prism. Equal volumes are produced due to the attributes of rectangular prisms.

## Student 3

Student 3 also attempts to create sample dimensions for one cylinder. He doubles and halves. He too seems to confuse prisms and cylinders, which may be reinforced by the dimensions he selects. This notion may be compounded by the dimensions he selects for his test cases.

## Student 4

Student 4 also creates a test case to compare the volumes. He shows the right ideas about doubling and halving. However, he incorrectly doubles the diameter to find the radius of the second cylinder. His correct conclusion is based on a calculation error.

*What would we want to ask these students? What might we do next?*

## Student 1

Since Student 1 makes no reference to how to find the volume of a cylinder, we should investigate if he understands this concept. If he can explain how the volume is found, we might have him create sample dimensions with friendly values to test if doubling and halving will produce equal volumes.

## Student 2

We must acknowledge that Student 2 has a strategy to test his theory. We must call attention to the incorrect formula. We should have him connect his work to the dimensions of the cylinder and prepare to ask about the circular face. We may also consider having him describe how prisms and cylinders are similar and different before revisiting the task.

## Student 3

Student 3's representation insinuates that the two will be equal, yet he overlooks the effect of squaring the radius. Let's assume he does know the formula. We may present "naked" values to establish why doubling and halving can't work. We can first ask if $(1)^2 = (2)^2$. From there, we can ask if $(1)^2 (4) = (2)^2 (4)$, leading him to consider why. With the latter established, we can revisit the prompt.

## Student 4

Student 4 has a reasonable strategy for testing his conjecture. However, due to a calculation error, it is unclear whether Student 4 truly understands why the volumes are not the same. It will be important to direct Student 4 to the incorrect radius and then ask him to reconsider. As a follow-up, we can explore why the volumes are not the same without using test dimensions.

## Student Work 1

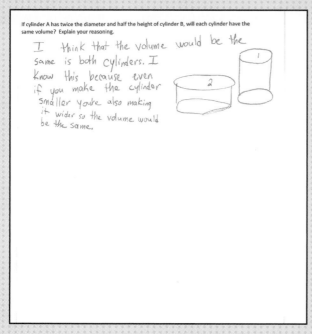

If cylinder A has twice the diameter and half the height of cylinder B, will each cylinder have the same volume? Explain your reasoning.

I think that the volume would be the same is both cylinders. I know this because even if you make the cylinder smaller you're also making it wider so the volume would be the same.

## Student Work 2

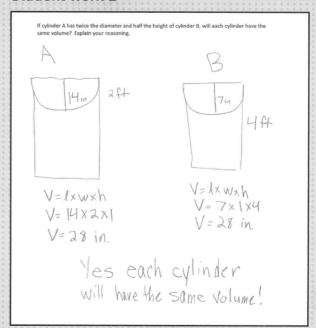

If cylinder A has twice the diameter and half the height of cylinder B, will each cylinder have the same volume? Explain your reasoning.

$V = l \times w \times h$
$V = 14 \times 2 \times 1$
$V = 28$ in.

$V = l \times w \times h$
$V = 7 \times 1 \times 4$
$V = 28$ in.

Yes each cylinder will have the same volume!

## Student Work 3

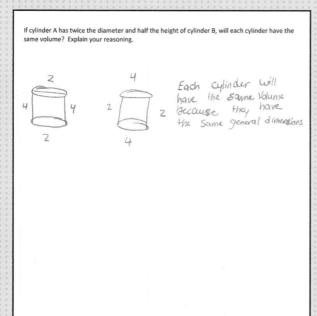

If cylinder A has twice the diameter and half the height of cylinder B, will each cylinder have the same volume? Explain your reasoning.

Each cylinder will have the same volume because they have the same general dimensions

## Student Work 4

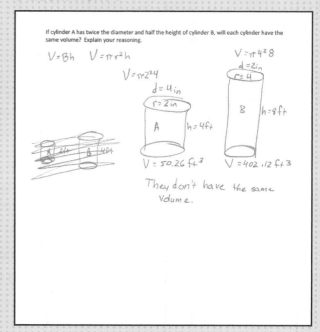

If cylinder A has twice the diameter and half the height of cylinder B, will each cylinder have the same volume? Explain your reasoning.

$V = Bh$  $V = \pi r^2 h$
$V = \pi 2^2 4$
$d = 4 in$
$r = 2 in$

$V = \pi 4^2 8$
$d = 2 in$
$r = 4$

$V = 50.26 ft^3$  $V = 402.12 ft^3$

They don't have the same volume.

# OTHER TASKS

- What will count as evidence of understanding?
- What misconceptions might you find?
- What will you do or how will you respond?

Visit this book's companion website at **resources.corwin.com/minethegap/6-8** for complete, downloadable versions of all tasks.

**TASK 36B:** Provide the dimensions for the cone displayed so that the volume of the cone is 48π cubic inches.

**Explain how you found these dimensions.**

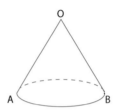

Finding the volume when given dimensions may not be evidence of full understanding of volume of cones. Task 36B reverses the traditional problem and will likely uncover a wide range of ideas and errors. To be successful, students will need to understand that the formula for the volume of a cone is $V = \frac{1}{3}\pi r^2 h$. It's likely that many students will apply trial and error to find the dimensions of the figure. Some students will look for factors of 48 and may mistakenly use these dimensions as the dimensions. For example, they might decide that the height is 8 inches and the radius is 6 inches. Some may connect ideas about triangles and divide 48 by 3 and find dimensions related to 16. Students who reason about the formula as an equation will recognize that they must determine factors of 144 (3 × 48) and that one of the factors is a perfect square since $r^2h = 144$. The perfect square factors of 144 are 4, 9, 16, and 36, so the possible radii are 2, 3, 4, or 6 inches. Considering the square root alone may be an oversight for some of our students with unfinished learning.

**TASK 36C:** A candy manufacturer is creating a new plastic case to hold small chocolate candies. They develop two prototypes. One prototype is a cone. The other is a cylinder.

The cone has a height of 4 inches and a diameter of 2 inches.

The cylinder has a height of 3 inches and a radius of 0.75 inches.

**Which prototype will hold the most candy? Justify your reasoning.**

Volume provides excellent opportunity for real-world problems such as the one in this task. Comparing different figures with different dimensions makes the problem both more challenging and more realistic. If the dimensions were the same, students would be able to reason that the cylinder will hold more candy. Many students will calculate the volume of each object and compare their results. Some students will realize that the cone has a larger radius by 0.25 inches and a larger height by 1 inch and make conclusions on this alone. *As we know, the cone is one-third the volume of the cylinder. These increased dimensions will still not compensate for the larger volume of the cylinder and thus the cone will still have a smaller volume.*

**MODIFYING THE TASK**

A potential extension could be to ask students how they could alter the dimensions of the cone so that the volume of the cone will be greater than the cylinder.

**TASK 36D:** **A cylinder holds 360π cubic centimeters. If you were to create a cone with the same volume, how would the dimensions of the cone compare to the dimensions of this cylinder?**

Cones, cylinders, and spheres have unique relationships as discovered by Archimedes. In this task, students demonstrate their understanding of the relationship between the volume of a cylinder and a cone. Students must first understand that a cone is one third of a cylinder with the same dimensions. Students must then recognize that to find a cone with the same volume, the height should be tripled. Some may invert the idea and divide by three. Some students may use trial and error, selecting dimensions of the cylinder to generate a volume of 360π cubic centimeters. Students may then try to alter the dimensions of the cone to find dimensions that also produce the same volume. Students with refined understanding will work from the relationship first to describe how to alter the dimensions.

NOTES

BIG IDEA

# 37

# BIG IDEA 37
## Angle Relationships

**TASK 37A**

Line EB is parallel to line JG. Find the measures for all of the missing angles.

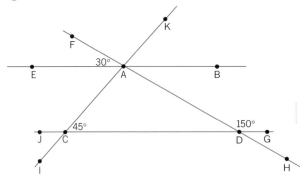

Note: This image is not drawn to scale.

**Explain how you found the measure of angle CAD.**

---

## About the Task

Understanding angle relationships is a foundation for the study of geometry. We can reason about the relationship between lines and intersections to determine unknown measures and other quantities. *This task synthesizes different ideas about angles and angle relationships in the same task instead of each missing angle representing the same idea.* Students will reason about angle relationships to find missing angles and determine if angles are congruent. Understanding of parallel lines and transversals is essential. Students will also need to apply their understanding of angles within a triangle to find missing angle CAD.

### PAUSE AND REFLECT

- How does this task compare to tasks I've used?
- What might my students do in this task?

Visit this book's companion website at **resources.corwin.com/minethegap/6-8** for complete, downloadable versions of all tasks.

## Anticipating Student Responses

Students will show a host of strategies for finding missing angles. In some cases, they will use ideas about supplemental angles. In others, they will rely on opposite angles. They may find angle CAD as the opposite of angle FAE, as a supplemental angle, or as the third measure of a triangle. There will be misconceptions as well. Some may assume that angle FAE is 90° by visual assumption. Students may be challenged to apply ideas of congruent angles formed by alternate interior angles as they have unfinished learning about them. Some may confuse complementary and supplementary angles, and others are likely to make calculation errors.

NOTES

### Student 1

Student 1 approximates one of the angles connected to point A by seeing that the angle is close to being a right angle. She disregards the note that says that the image is not drawn to scale. Student 1 states that "C and D are next to each other making them the same." It seems that she believes that triangle CAD is an isosceles triangle. While Student 1 makes some inappropriate assumptions, she is able to correctly apply understanding of vertical angles.

### Student 2

Student 2 makes an initial error by calculating angle ADC as 40° rather than 30° to form a linear pair. This impacts her calculations for the remainder of the angle measures. Despite this mistake, Student 2 shows an understanding of the relationship between vertical angles as well as between the angles in a triangle. Student 2 explains, "I found the angle CAD by finding the angle at every line till they added up to 180."

## USING EVIDENCE

*What would we want to ask these students? What might we do next?*

### Student 1

First, it is important to acknowledge that Student 1 understands that the measures of angles of a triangle have a sum of 180° and is able to find the measure of vertical angles. It is also necessary to emphasize that because the diagram is not drawn to scale, we cannot assume any angle measure because the angle appears to be a certain measure. Since Student 1 doesn't demonstrate understanding of supplementary angles, we need to confirm that she does understand them and what the relationship between these angles is. We can then have her identify the supplementary angles in the diagram.

### Student 2

Like Student 1, Student 2 needs additional work to reinforce the relationship between supplementary angles. This should help clarify if the student made a calculation error or has a true misconception about supplementary angles. We might then redirect her to confirm her calculations. We might further develop her strategies for finding missing angles by discussing what is known about triangles due to triangle ACD being part of the diagram. We might facilitate a discussion to highlight that finding the interior angles of the triangle can be used as a strategy for finding angles at the top of the diagram.

## TASK 37A: Line EB is parallel to line JG. Find the measures for all of the missing angles.

Explain how you found the measure of angle CAD.

## Student Work 1

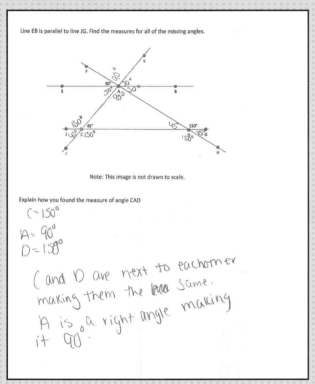

Line EB is parallel to line JG. Find the measures for all of the missing angles.

Note: This image is not drawn to scale.

Explain how you found the measure of angle CAD

C = 150°
A = 90°
D = 150°

C and D are next to eachother making them the same.
A is a right angle making it 90°.

## Student Work 2

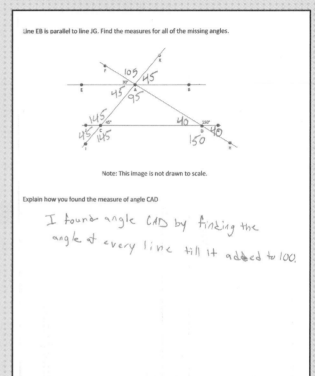

Line EB is parallel to line JG. Find the measures for all of the missing angles.

Note: This image is not drawn to scale.

Explain how you found the measure of angle CAD

I found angle CAD by finding the angle at every line till it added to 100.

## Student 3

Student 3 correctly identifies vertical angles in the diagram. She also correctly finds the measure of angle CAD, apparently using the sums of the measures of the angles of a triangle. Student 3, however, is unable to find any of the other missing values. She also has numbers, such as 40, 10, 50, and 65, next to each point on the graph, but it is unclear why these measures are listed. For the explanation, Student 3 states, "I used my brain to figure out CAD."

## Student 4

Student 4 correctly identifies each of the angles in the diagram. Student 4 explains, "I know that the angle CDA is 30° and a triangle's angles add up to 180°, so 180 − (45 + 30) = 105." Though she doesn't answer the prompt specifically, we can see that her logic is accurate and can be transferred to the other angle.

### *What would we want to ask these students? What might we do next?*

## Student 3

We can build on Student 3's understanding of angle sums within a triangle and that she is able to find the measure of vertical angles. We need to understand why she writes these measures next to each point rather than adjacent to a specific angle. There may be potential misconceptions, though it is difficult to determine from her labels or explanation. We can prompt her to find specific angles such as angle JCA. This will also reveal whether Student 3 understands the relationship between supplementary angles. As with Students 1 and 2, Student 3 also needs additional work developing clear and complete explanations. Something as simple as "CAD is supplemental to angles EAB and BAD, and supplemental angles have a sum of 180°" can suffice.

## Student 4

Student 4 appears to understand how to use vertical angles, supplementary angles, and the sum of measures of angles of a triangle to find missing angle measures. Her explanation cites the wrong angle, but her reasoning is sound. We may prompt her to identify if there is another way to find the measure of these angles of the triangle. We can advance her understanding by investigating different diagrams that may feature more than two parallel lines, complementary angles, or different polygons.

**TASK 37A:** Line EB is parallel to line JG. Find the measures for all of the missing angles.

Explain how you found the measure of angle CAD.

## Student Work 3

Line EB is parallel to line JG. Find the measures for all of the missing angles.

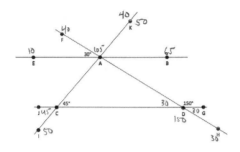

Note: This image is not drawn to scale.

Explain how you found the measure of angle CAD

I used my brain to figure CAD.

## Student Work 4

Line EB is parallel to line JG. Find the measures for all of the missing angles.

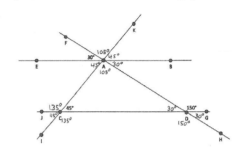

Note: This image is not drawn to scale.

Explain how you found the measure of angle CAD

I know that angle CDA is 30° and a triangle's angles add up to 180°, so 180-(45+30)=105.

## OTHER TASKS

- What will count as evidence of understanding?
- What misconceptions might you find?
- What will you do or how will you respond?

**TASK 37B:** Identify two sets of possible angle measures for triangle ABC.

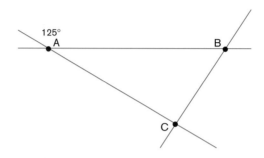

**Is it reasonable for angle B to have a measure of 70 degrees? Why or why not?**

This task examines students' understanding of supplementary angles as well as the sum of angles within a triangle. Students will use this to determine the measure of angle BAC to be 55°. *Then students must find possible pairs of angles that have a sum of 125° for the remaining two angles within the triangle.* Some students may note results based on an assumption that BC is perpendicular to AC or that angle C is 90°, though that is not given. As we know, there are many possibilities. We should look for students who find one solution set, others who find one solution but note that there are more possibilities, and still others who find multiple solution sets. Yet, in all cases, angle C should have a greater measure than angle B. With this in mind, students are asked to consider a reasonable measure of angle B. If that was the case, C would have a value of 55° because we know angle A to be 55° and the sum of a triangle to be 180°. Students may disregard the relative size of the angles and falsely assume that this is a viable solution for angle B.

**MINING TIP**

Precision means that students label angles correctly. We should make sure students are labeling which angle in the pair corresponds to angle B and which corresponds to angle C. Check to see that students have labeled C as the greater angle.

**TASK 37C:** Find the exterior angles of triangle ABC.

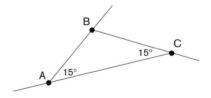

**Explain how you found the angle measures.**

Task 37C also requires our students to apply understanding of supplementary angles and the sum of angles in a triangle. However, in this task, students must first find angles within a triangle in order to find the exterior angle B. Successful students will recognize that angles A and C are congruent and so their exterior angles must both be 165°. These students may realize that the third exterior angle may be found by subtracting the other exterior angles from 360°. We may find students who hold the misconception that the sum of exterior angles is also 180° or those that exchange the sum of interior angles with exterior. Though we can restate this often, it is likely more effective to have them use full circle protractors and explore exterior angle sums for a variety of triangles. We might also connect the relationships between interior and exterior angles and discuss how a set of knowns can help us find unknowns.

**TASK 37D:** In the figure, lines AB, CD, and EF are parallel.

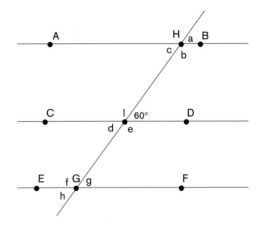

## Which of the following angles (a-h) would *not* have a measure of 60 degrees? How do you know?

Students will need to apply understanding of parallel lines cut by a transversal to identify congruent angles. Students will also need to identify pairs of vertical angles, alternate interior angles, and corresponding angles. Students may have difficulty translating with three parallel lines. Students also typically confuse alternate interior angles with supplementary angles. *We should also look for students who visually compare angle measures rather than use angle proofs.* We may consider extending the task by adding a fourth parallel line or by removing any measurements and ask students to generate possible solution sets.

**MINING HAZARD**

**Students often work with diagrams in which parallel lines are featured horizontally or vertically. We may consider rotating the lines AB, CD, and EF to investigate student perception and reasoning.**

NOTES

# BIG IDEA 38
## Transformations, Similarity, and Congruence

### TASK 38A

Create a quadrilateral in the coordinate plane.

Perform two transformations on the image.

Describe how each transformation mapped the new image.

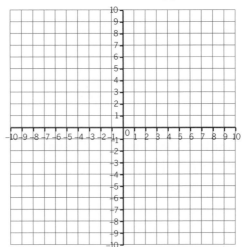

### About the Task

Transformation is a significant concept that connects to both geometric and algebraic contexts. Students need to build fluency with identifying and applying transformations in and out of the coordinate plane. As students work with transformations, they begin to differentiate between transformations that

## PAUSE AND REFLECT

• How does this task compare to tasks I've used?

• What might my students do in this task?

Visit this book's companion website at
**resources.corwin.com/minethegap/6-8**
for complete, downloadable versions of all tasks.

generate similar figures and transformations that produce congruent figures. For this task, students create a quadrilateral in the coordinate plane and then must apply two transformations and describe how these transformations map to the new image. As students apply their transformations, it is critical to pay attention to how they are labeling the new vertices of the transformed images to ensure that they are accurately applying the transformation.

## Anticipating Student Responses

Students will begin by drawing a quadrilateral somewhere in the coordinate plane. Some students may neglect the directions and draw a figure that is not a quadrilateral. Our students are likely to apply different types of transformations to the figure. The most common transformations will be reflections, translations (slides), or rotations. It's likely that some students will apply the same transformation twice. Some students may attempt to produce a dilation from the center or from one of the vertices. Some students will revisit the original image for the second transformation, while others will transform the newly generated transformed image. Students may accurately transform the image but inaccurately label the image to represent a different type of transformation.

NOTES

### Student 1

Student 1 fails to create a quadrilateral in the coordinate plane. Instead, he creates three points in the plane at (0, 0), (1, 1), and (2, 2). Student 1 incorrectly identifies the coordinate as (2, 3). It appears that Student 1 applies a reflection over the $x$-axis to produce the points (0, 0), (1, –1), and (2, –2). Student 1 doesn't provide any explanation or description of the transformation performed.

### Student 2

Student 2 doesn't produce a quadrilateral or label the vertices. Student 2 explains that he performed a reflection and a rotation. His representations appear to represent a reflection over the $y$-axis and a rotation about the origin.

## USING EVIDENCE

***What would we want to ask these students? What might we do next?***

### Student 1

While Student 1 does not provide much work, it appears that he may have some basic understanding of a reflection. We should be sure that he knows what a quadrilateral is and that he can draw and label its vertices. We want to determine what he knows about the possible transformations and that he can model them. We may first have him work with single transformations of a quadrilateral before adding additional transformations. Physical models placed on the coordinate grid or tracing paper may also help him see the impact of these transformations.

### Student 2

We can apply similar strategies with Student 2. Again, we want to confirm understanding of quadrilaterals. We should also confirm understanding of labeling. We want to clarify which transformation is represented with each new image. We can provide physical models for him to use at his discretion. After establishing understanding of these transformations and translations, we can challenge him to apply more than one transformation to a shape. As understanding of multiple transformations develops, we can begin to connect these ideas to shapes represented on a coordinate grid.

## Student Work 1

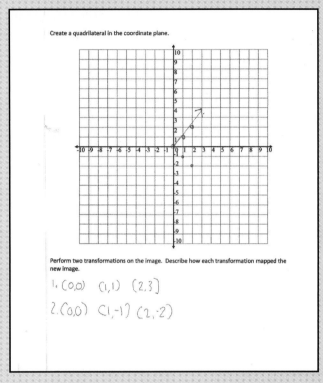

Create a quadrilateral in the coordinate plane.

Perform two transformations on the image. Describe how each transformation mapped the new image.

1. (0,0)  (1,1)  (2,3)

2. (0,0)  (1,-1)  (2,-2)

## Student Work 2

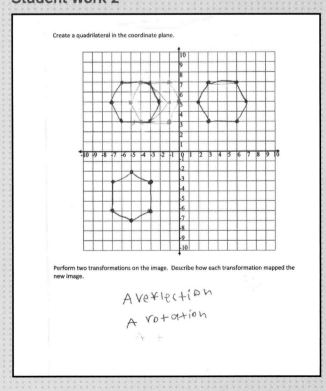

Create a quadrilateral in the coordinate plane.

Perform two transformations on the image. Describe how each transformation mapped the new image.

A veflection

A rotation

## Student 3

Student 3 draws and labels the vertices of a quadrilateral. He appears to apply a clockwise rotation about the origin. It also appears that he may have performed a translation of the original image. Student 3 has difficulty describing the transformations. He states, "Some went up and down and others went side to side." He doesn't identify which transformations were specifically applied.

## Student 4

Like Student 3, Student 4 accurately draws and labels the vertices of a rectangle. The student writes that the one set of images represents a "flip over the *y*-axis." Student 3 also describes that he performed a 90° counterclockwise rotation. It is interesting that he labels the points in his reflected image but not in his rotated image. It may signal less understanding than what we might first assume.

### *What would we want to ask these students? What might we do next?*

## Student 3

Student 3 shows different understanding and misunderstanding than Students 1 and 2. He shows a quadrilateral and applies transformations to it. He can describe what happens in very raw ways. Our work is to help him develop the mathematics vocabulary to accurately and fully describe what is happening in his work. We even begin with less complicated and more ordinary terms to support the development of precise vocabulary. For example, he might use "flip" to develop reflection or "turn" to develop rotation. As his vocabulary grows, we can introduce new, but related, terms and concepts. In this case, we can have him assess if the transformed image is similar or congruent to the original image.

## Student 4

Student 4 understands the transformations he performed to some extent. His missing labels of the rotated image may indicate that he can visualize how the image of the shape changes but isn't sure about the result of the specific corners of the quadrilateral. He may need additional work to explore how the relative location of sides or vertices changes as they are transformed. We can support this development by cutting out shapes of clear plastic and labeling their points. After transformations are applied to the clear shapes, Student 4 can observe and record the new locations of these points. As with Student 3, we can have Student 4 consider if the images are similar or congruent. We can also begin to have these students work with applications of multiple transformations to a shape or figure.

**TASK 38A:** Create a quadrilateral in the coordinate plane. Perform two transformations on the image. Describe how each transformation mapped the new image.

### Student Work 3

Create a quadrilateral in the coordinate plane.

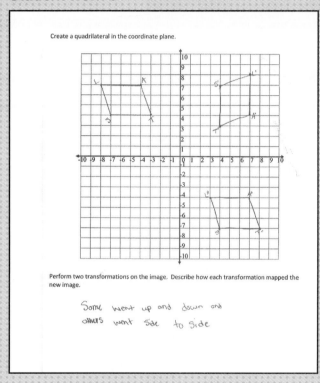

Perform two transformations on the image. Describe how each transformation mapped the new image.

Some went up and down and others went side to side

### Student Work 4

Create a quadrilateral in the coordinate plane.

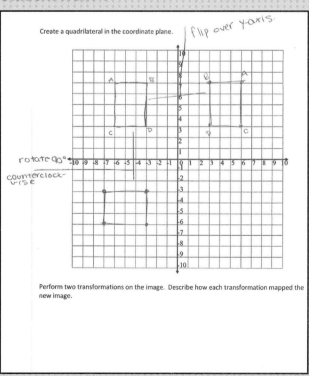

Perform two transformations on the image. Describe how each transformation mapped the new image.

## oTHER TASKS

- What will count as evidence of understanding?
- What misconceptions might you find?
- What will you do or how will you respond?

 Visit this book's companion website at **resources.corwin.com/ minethegap/6-8** for complete, downloadable versions of all tasks.

**TASK 38B:** Describe the transformation or set of transformations that maps triangle ABC onto triangle A′B′C′.

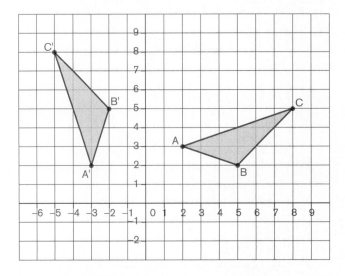

 **MINING TIP**

**Students need to differentiate between transformations that produce similar figures and ones that produce congruent figures. Each transformation will produce a similar figure. A congruent figure is a special case of similar figures in which the ratio of the sides is 1:1.**

This task requires students to identify the transformation or set of transformations that maps the triangle onto its transformed image. Students may try to use guess and check by selecting a transformation and then testing the points. Most students will recognize that this represents a rotation, but students with refined understanding will articulate that this is a 90° counterclockwise rotation about the origin. For students having trouble recognizing transformations, physical models, clear plastic models (as noted with Student 3 in task 38A), and tracing paper are useful supports. We can have students trace the original triangle (in this case) and vertices and then model how to fold or rotate the tracing paper to test or produce different transformations. *Next, have students identify whether the two images are similar or congruent.* As students use these tools, it is critical that we connect the physical model with the representation with the task or problem.

**TASK 38C:** Quadrilateral ABCD is shown on the coordinate grid.

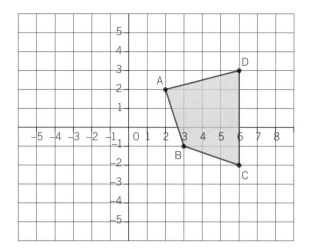

Perform two of the following transformations on ABCD and describe how the transformed image compares to the original image.

- **Reflection**
- **180° rotation**
- **90° rotation**
- **Dilation**
- **Translation**

This task is another opportunity to apply understanding of transformations of a quadrilateral. Unlike task 38A, this task provides specific transformations, including a dilation. Students will select the transformations that they are most comfortable modeling. Their choice may also indicate the transformations that we have featured more prominently in our instruction. We should pay attention to the labels of the vertices of the transformed image as well as the clarity with which the student can describe the comparison between images. Students may be able to manipulate the image but be challenged to label the vertices accordingly.

**TASK 38D:** Sheila performed a transformation on a triangle. She produced an image that has preserved the angle measures of the image.

**What transformation might she have performed? Why?**

Task 38D examines students' understanding of which transformations preserve angle measure and which preserve distance between vertices. Since the task states that angle measure is preserved, students may select any of the transformations. Some students may select a dilation because only angle measure was specifically referenced. Other students may select one of the transformations that produces congruent figures. When working with these students, we should ask what other attributes are preserved with their specified transformation.

# BIG IDEA 39
## Distance and the Pythagorean Theorem

### TASK 39A

**Find the length of each segment in figure ABCDE.**

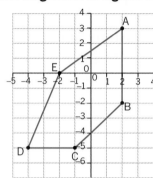

**Explain your reasoning.**

## About the Task

Finding the distance between two points has many applications in algebra and geometry. Around sixth grade, students learn to find horizontal and vertical distances between points in the coordinate plane. Around eighth grade, students expand this understanding to find the distance between any two points in the coordinate plane. Students learn to use and apply the Pythagorean Theorem in order to find lengths of segments. In this task, students apply this knowledge to find the length of each segment of a figure in the coordinate plane.

### PAUSE AND REFLECT

- How does this task compare to tasks I've used?

- What might my students do in this task?

 Visit this book's companion website at **resources.corwin.com/minethegap/6-8** for complete, downloadable versions of all tasks.

## Anticipating Student Responses

Some students may mistakenly count spaces to find the length of each segment, including the diagonals. Other students will likely approach this problem in one of two ways. They may apply the Pythagorean Theorem to find the lengths of segments. These students may forget to square quantities or use the square root of the sum to find the distance. *Other students may apply the distance formula* $d = \sqrt{(x_2 - x_1)^2 + (y_2 - y_1)^2}$. Students using the distance formula may be more likely to make computational mistakes or interchange values, such as subtracting an *x*-value from a *y*-value.

⚠ **MINING HAZARD**

Students who apply the formula may not understand why it works. Without understanding, retention and application of ideas are less likely. Also, the lack of understanding challenges our students' ability to determine the reasonableness of their solutions.

NOTES

## Student 1

Student 1 attempts to approximate the length of each segment by counting the boxes. While this strategy works for the horizontal and vertical line segments, it is not an effective strategy for the diagonal line segments. For example, Student 1 determined that the length of segment BC is 3 units because the segment passes through 3 square units. She overlooks the fact that the length of a diagonal within a square is longer than the length of a side. Student 1 makes no connection between this problem and the Pythagorean Theorem.

## Student 2

Student 2 confuses finding the length of segments with the area of regions. It appears she is trying to find the area of the composite figure. This is confirmed in her description and diagram. She states that she "got 24 by making each triangle into squares then finding the number of blocks on each side and then dividing by 2." Student 2 also makes some counting errors as she measures the horizontal and vertical lengths. Student 2 does correctly find the length of CD. Student 2 also makes no reference to the Pythagorean Theorem.

*What would we want to ask these students? What might we do next?*

## Student 1

We can build on Student 1's ability to identify the length of segments AB and CD. She must understand that the length of a diagonal is a measurement different from counting spaces or, at the very least, longer than the length of sides. We might have her create a similar model on a peg or geoboard. She can use string or something similar to find the length of a vertical or horizontal side and compare that to the diagonal. Or, we can have her determine the length of the diagonal using her strategy and then compare that value using the string. We might have her use a centimeter ruler or something similar to compare measurements as well. It is most important to position her so that she can find flaws in her approaches. We can leverage her dissonance to introduce or reinforce appropriate strategies.

## Student 2

Student 2 misunderstands the purpose of the task. She notices a composite figure and assumes that the task intends for her to find the area of the figure. We might first ask her to reread the directions and explain how her response relates to the length of each segment in the figure. It's possible that she considers perimeter and area to be interchangeable. If so, our next steps will be to revisit both concepts with simple polygons. She may realize that she found area rather than distance. In this case, we can highlight that she has already created triangles on the diagram. We can ask how that can help her find the perimeter of the figure. At this point, we'll need to look for her to use the Pythagorean Theorem, the distance formula, or a strategy similar to that of Student 1.

## TASK 39A: Find the length of each segment in figure ABCDE.

Explain your reasoning.

## Student Work 1

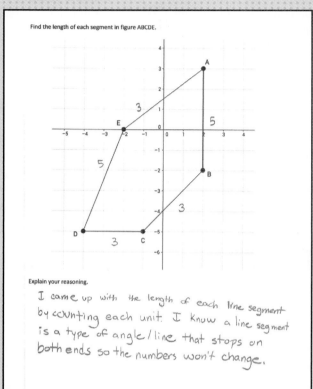

**Find the length of each segment in figure ABCDE.**

**Explain your reasoning.**

I came up with the length of each line segment by counting each unit. I know a line segment is a type of angle/line that stops on both ends so the numbers won't change.

## Student Work 2

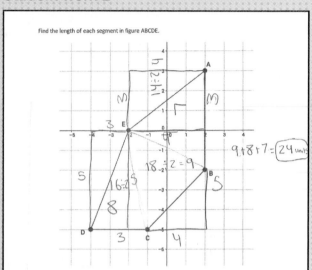

**Find the length of each segment in figure ABCDE.**

**Explain your reasoning.**

I got 24 by making each triangle into squares then finding the number of blocks on each side and then dividing by 2. I did this because when you go from a triangle to square it doubles in size so you would divide by two to get the amount of units for a triangle.

### Student 3

Student 3 recognizes that right triangles may be used but is unable to apply the Pythagorean Theorem correctly. She does accurately create reference triangles with the hypotenuse as the segment of interest. Student 3 correctly finds the lengths of the horizontal segment CD and vertical segment AB. For the diagonal line segments, she correctly identifies the lengths of the sides in her triangles but simply adds these distances to generate the length of the hypotenuse.

### Student 4

Student 4 recognizes that the Pythagorean Theorem should be applied to find the lengths of the diagonals. She accurately finds the lengths of the segments CD and AB. She creates triangles for each of the diagonal segments. She explains "I used pathaguras theorem to find the missing length of the line and then I used $a2 + b2 = c2$ to get my answer." Student 4 uses decimal approximations for those two segments.

*What would we want to ask these students? What might we do next?*

### Student 3

We must acknowledge that Student 3 uses an appropriate strategy to find the length of the horizontal and vertical line segments. She also has the right ideas about how to find the values of the diagonal line segments. But, she doesn't appear to understand the relationship between the sides of a right triangle. As with Student 1, we want to be sure she understands why these values cannot be correct. We might do this by having her mark the length of the hypotenuse on a small strip of paper and compare that to the sum of the lengths of the other two sides. She should see that she has overestimated the length of the hypotenuse. Once she realizes this, we can engage her in a review of the Pythagorean Theorem and have her apply it to find the length of each remaining segment.

**CONTENT CONNECTION**

Applying the Pythagorean Theorem provides an opportunity for students to connect with the work they do with irrational numbers and square roots in Grade 8. This task is an opportunity to reinforce representing irrational numbers with square roots as well as strategies for approximating square roots.

### Student 4

Student 4 understands that in order to find the length of each diagonal segment, she can apply the Pythagorean Theorem. She finds a decimal approximation for the length of some of the diagonal segments. A next step with her is to engage her in reflection about the accuracy of her answers. While her approximations are fairly close to the answer, we can ask how she could represent the length with an exact solution. *We can help connect her understanding of square roots to represent irrational numbers.*

**TASK 39A:** Find the length of each segment in figure ABCDE.

Explain your reasoning.

## Student Work 3

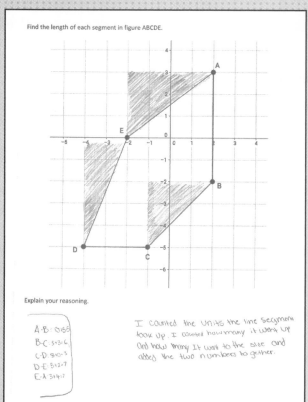

Find the length of each segment in figure ABCDE.

Explain your reasoning.

A-B: 0·1·5·5
B-C: 3+3·6,
C-D: 3+0·3
D-E: 5+2·7
E-A: 3+4·7

I counted the Units the line segment took up. I counted how many it went up and how many it went to the side and added the two numbers to gether.

## Student Work 4

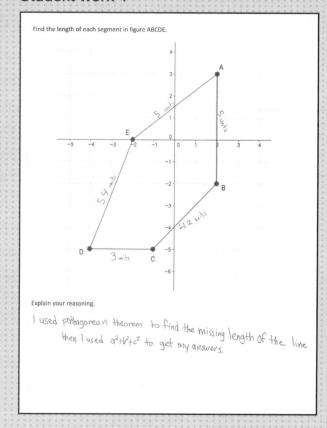

Find the length of each segment in figure ABCDE.

Explain your reasoning.

I used pathagorean theorem to find the missing length of the line then I used $a^2+b^2=c^2$ to get my answers.

## OTHER TASKS

- What will count as evidence of understanding?
- What misconceptions might you find?
- What will you do or how will you respond?

 Visit this book's companion website at **resources.corwin.com/ minethegap/6-8** for complete, downloadable versions of all tasks.

**TASK 39B:** Which of the following triangles are right triangles?

Side lengths: 3 cm, 4 cm, 5 cm

Side lengths: 5 cm, 6 cm, 7 cm

Side lengths: 5 cm, 12 cm, 13 cm

Side lengths: 9 cm, 12 cm, 15 cm

Side lengths: 10 cm, 13 cm, 16 cm

**Justify that one of your selections is or is not a right triangle.**

**Find the dimensions of another right triangle.**

**CONTENT CONNECTION**

**This task is an extension of students' understanding of dilations and proportional figures. We can use this opportunity to build connections between special right triangles and proportionality of right triangles.**

The lengths of right triangles have a unique relationship. In this task, students consider triangles that represent this relationship. Students may choose to test each set of side lengths using the Pythagorean Theorem. The first and third choices are two examples of right triangles. *Some students will recognize the fourth triangle (9-12-15) as a right triangle because it is a dilation of choice A with a scale factor of 3.* Students' misconceptions about this relationship may cause them to select the second triangle (5-6-7) as it relates directly to the right triangle above (3-4-5). The same type of error is replicated in the fifth triangle (10-13-16), which relates directly to the fourth triangle (9-12-15).

**TASK 39C:** Triangle 1 is a right triangle with one side length of 6 inches. Triangle 2 is similar to triangle 1 and has one side length of 9 inches. Sketch a possible example of triangle 1 and triangle 2 and label all side lengths.

Task 39C makes connections between right triangles and similar figures. Students are likely to draw right triangles that are similar before finding missing side lengths. Students will need to determine another side length for one of the triangles and then use the Pythagorean Theorem to approximate the length of the third side. Our students will need to use proportional reasoning to find the lengths for the similar triangle. Some students may choose one triangle to have lengths 6 inches, 8 inches, and 10 inches, since this is a dilation of a 3-4-5 right triangle. Students who are still developing their understanding of the Pythagorean Theorem or similar figures may be challenged by these examples.

**TASK 39D:** Mr. Novak needs to clean out the gutters on his house. The gutters sit on the edge of the house, 15 feet above the ground. There is a bush along the front edge of the house, so Mr. Novak will have to place a ladder 5 feet from the house.

**What length of ladder will Mr. Novak need to reach the gutters?**

This task provides a real-world context for students to apply the Pythagorean Theorem. Students may misread the problem and assume that the length of the ladder is 15 feet, causing them to subtract the squares of the quantities rather than finding the sum. Students may forget to square the side lengths or take the square root of the sum. *We can support student understanding by encouraging students to draw a diagram to represent the problem.* Students will find a hypotenuse of 15.81 feet. This solution in context provides an excellent opportunity for discussion about actual lengths of ladders and how these distances change as a ladder is moved toward or away from the house.

**MINING TIP**

Diagrams can be useful to help understand the problem. In this example, the diagram will help students see that they must find the hypotenuse of the right triangle.

NOTES

# CHAPTER 7

# STATISTICS

# BIG IDEA 40
## Univariate Categorical Data

**BIG IDEA**
**40**

### TASK 40A

Jaime is doing research on siblings. She samples 50 students in Grade 6 and 50 students in Grade 7 to see how many students have siblings. Their responses are shown in the bar graphs below.

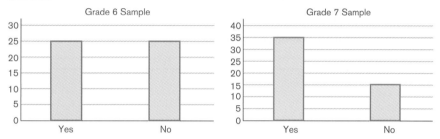

Grade 6 Sample

Grade 7 Sample

For which group is the variation in the responses smallest? Justify your answer.

### About the Task

As our students begin to develop an understanding of statistics, they learn to differentiate between categorical and quantitative data. Each type of data provides vital information, yet the ways that we represent and analyze these different data types must be differentiated. A common misconception is that because categorical data represent frequency, it is not possible to assess variation between the categorical variables. The purpose of this task is to address that specific misconception. When working with variation among categorical variables, we must answer the question, "If another person was surveyed, how confident would we be in predicting his or her response?"

### PAUSE AND REFLECT

- How does this task compare to tasks I've used?

- What might my students do in this task?

 Visit this book's companion website at **resources.corwin.com/minethegap/6-8** for complete, downloadable versions of all tasks.

The more confident we are in our prediction, the less variation that exists between the variables. In this task, students are to determine which group of responses has the smallest variation.

## Anticipating Student Responses

Since there are only two groups to analyze, students will select either the Grade 6 sample or the Grade 7 sample. Because of this, it will be critical to carefully analyze student justifications for their selection. A common mistake is to select Grade 6 because the graph appears to have similar totals. These students think about the problem as they would with quantitative data or look for clustering of data. For quantitative data, a clustering of data means less variation. Successful students will understand that the smallest variation is for the Grade 7 sample, because more students responded yes. If another Grade 7 student is surveyed, there is a higher likelihood that he or she will respond yes. However, with categorical variables, variation does not follow this same logic.

NOTES

### Student 1

Student 1 shows that the sum of each sample is equal to 50 students. He concludes, "I think none of them have the smallest response because the number of responses are the same." Student 1 associates variation with total number of responses rather than associating variation with comparing the responses for each category.

### Student 2

Student 2 makes the most common error for this type of task. He uses the definition of variation for quantitative data and applies it to categorical data. Student 2 explains his reasoning for selecting Grade 6 by stating, "I know this because the variation I'm pretty sure is the distance between two answers, and 6th grade has a smaller one."

## USING EVIDENCE

*What would we want to ask these students? What might we do next?*

### Student 1

When working with Student 1, we might first have him describe the two graphs. We can ask, "How are the graphs similar and how are they different?" We should emphasize that since both samples contain 50 responses, this makes the comparison easier because we can look at counts rather than percents of responses. He should be asked to describe how he understands variation. We can relate his definition to the concept of categorical data. We may need to explain that a smaller variation means that it is easier to predict what another student sampled may select. We will need to provide more opportunities to apply his understanding. As we do this, we should be sure to use familiar contexts and rather predictable results so he can reinforce his learning.

### Student 2

Our next step for Student 2 is obvious. He needs to distinguish between the behaviors of quantitative data and categorical data. The best way to address this is to think about predictability of another response. We can ask him to predict and explain what the response of the fifty-first student is in each grade. His explanation will provide insight into how he perceives the data. It will also help us determine next steps specifically. In either case, we should continue to work with examples for quantitative and categorical data to help Student 2 develop understanding of the difference. After his learning advances, we may have him revisit his work in this task to determine how his thinking has changed.

**TASK 40A:** Jaime is doing research on siblings. She samples 50 students in Grade 6 and 50 students in Grade 7 to see how many students have siblings. Their responses are shown in the bar graphs.

For which group is the variation in the responses smallest? Justify your answer.

## Student Work 1

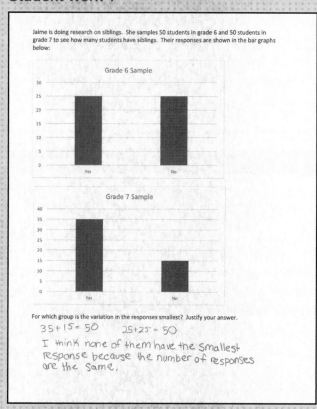

Jaime is doing research on siblings. She samples 50 students in grade 6 and 50 students in grade 7 to see how many students have siblings. Their responses are shown in the bar graphs below:

For which group is the variation in the responses smallest? Justify your answer.

$35 + 15 = 50$    $25 + 25 = 50$

I think none of them have the smallest response because the number of responses are the same.

## Student Work 2

Jaime is doing research on siblings. She samples 50 students in grade 6 and 50 students in grade 7 to see how many students have siblings. Their responses are shown in the bar graphs below:

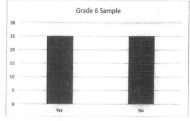

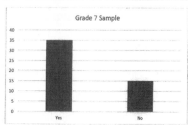

For which group is the variation in the responses smallest? Justify your answer.

The 6th grade sample has a smaller variation. I know this because variation I'm pretty sure is the distance between two answers, and 6th grade has a smaller one!

### Student 3

Student 3 incorrectly identifies the Grade 6 sample as having the smallest variation. He selects Grade 6 because "25 students have siblings and 25 don't have siblings. It's a 50% chance." Interestingly, Student 3 considers likelihood of responses for each category in Grade 6.

### Student 4

Student 4 correctly identifies Grade 7 as the sample with the smallest variation. He explains, "I think the grade 7 response because grade 6 is evenly distributed 25 to 25 but grade 7 is 35 to 15." While he makes the correct selection and begins to describe the difference in number of responses, his justification does not provide enough insight as to whether he understands why Grade 7 has a smaller variation.

*What would we want to ask these students? What might we do next?*

### Student 3

For Student 3, we can build on his thinking of likelihood of different responses. We can ask him what the probability of each category would be for Grade 7. He should be able to describe the likelihood even if he is unable to do so with exact values. We can have him compare each of the probabilities. It may be as easy as explaining to him that smaller variation exists for categorical data when there is a greater chance of one outcome. We can have him revisit his work to see if he uses this new idea to modify his response. Like the others, Student 3 needs additional opportunities to work with variation. We might continue to leverage his thinking about likelihood as he works with these new tasks.

### Student 4

We need to check to make sure that Student 4 understands why the Grade 7 sample has a smaller variation. We can ask him to explain why the 35 to 15 responses would indicate that the variation is smaller. We might consider having students like Students 3 and 4 review each other's ideas to create a new collaborative justification. Doing so helps each student refine his or her thinking.

**TASK 40A:** Jaime is doing research on siblings. She samples 50 students in Grade 6 and 50 students in Grade 7 to see how many students have siblings. Their responses are shown in the bar graphs.

For which group is the variation in the responses smallest? Justify your answer.

## Student Work 3

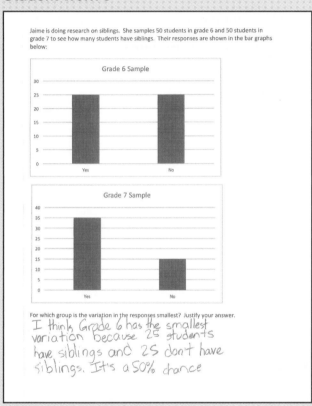

Jaime is doing research on siblings. She samples 50 students in grade 6 and 50 students in grade 7 to see how many students have siblings. Their responses are shown in the bar graphs below:

For which group is the variation in the responses smallest? Justify your answer.

I think Grade 6 has the smallest variation because 25 students have siblings and 25 don't have siblings. It's a 50% chance

## Student Work 4

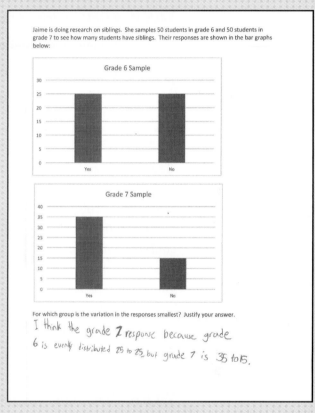

Jaime is doing research on siblings. She samples 50 students in grade 6 and 50 students in grade 7 to see how many students have siblings. Their responses are shown in the bar graphs below:

For which group is the variation in the responses smallest? Justify your answer.

I think the grade 7 response because grade 6 is evenly distributed 25 to 25, but grade 7 is 35 to 15.

Visit this book's companion website at **resources.corwin.com/ minethegap/6-8** for complete, downloadable versions of all tasks.

**TASK 40B:** Which of the following represents categorical data? Why?

- An online store gathers data on customer zip codes

- A new employer does research on typical salaries at a similar company

- A teacher surveys her classes to determine how many minutes students study

- A pet store surveys customers to determine favorite type of pet

**MINING HAZARD**

Students often overgeneralize data with numbers to always represent quantitative data. Examples like the zip code scenario are critical to help students differentiate between categorical and quantitative data.

Task 40B can help determine if students can differentiate between categorical and quantitative data. Some students will correctly identify the zip codes and pet examples as categorical. The other examples represent quantitative data. *A common misconception is for students to ignore the zip codes example because zip codes contain numbers.* To address this misconception, we should revisit their basic understanding of categorical and quantitative data. Categorical data are used to sort into categories and find frequencies, whereas quantitative data are used to analyze shape, center, and spread. For quantitative data, finding the mean and median provides a relevant insight into the data. For the zip codes example, the mean or median zip code would not provide any useful information about the data.

**TASK 40C:** Mrs. Wiegmann is taking her class to the petting zoo. She asks her students to identify their favorite farm animal. Their responses are displayed below.

| sheep | cow | cow | sheep | horse | horse | horse |
|-------|---------|------|-------|-------|-------|-------|
| pig | chicken | goat | horse | goat | horse | goat |
| cow | sheep | cow | cow | horse | goat | |

**Create an appropriate display to summarize the responses.**

For this task, students must analyze the categorical data and create an appropriate display. The task purposefully does not state which display to create, so students must determine which is the best display for the data. For categorical data, students should be selecting among bar graphs and circle graphs (pie charts). Bar graphs are used to compare relative frequency of each categorical variable, whereas circle graphs display what percent of the whole each category constitutes. If students select a bar graph, they need to use a relative scale that has includes a frequency scale of at least 6 units. If students select a pie chart, they must accurately calculate percentages out of 20 samples. The circle graph should be divided into the following sections: 15% sheep, 25% cow, 30% horse, 5% pig, 5% chicken, and 20% goat. *Students may have difficulty dividing the circle into the appropriately sized sections.* If necessary, we can have students consider which of the percentages are the easiest to display and then work to show the more challenging percentages.

**CONTENT CONNECTION**

This task provides an opportunity to connect understanding of creating pie charts to fraction and percent understanding. Students need to think of the circle as the whole and need to subdivide based on 20 equal parts.

**TASK 40D:** Elyse is planning invitations to a pool party. She wants to make a dessert for her friends to enjoy. The school recently took a survey of favorite desserts. The survey data are displayed in the bar graph.

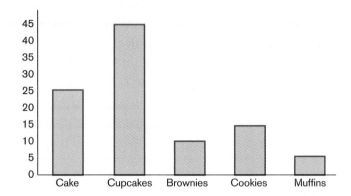

**Convert the bar graph to a circle graph and summarize the data.**

In this task, students convert between one representation of categorical data to another. To successfully convert this graph, students must consider the total sample size of 100 students. Once students recognize that each of these frequencies is out of 100, they should be able to convert to a circle graph. Some students may struggle with representing the 5%. For these students, discuss the relative size of the region that is 10% compared to 5%.

NOTES

# BIG IDEA 41
## Univariate Quantitative Data

### TASK 41A

> Mrs. Gittermann gave a quiz to each of her three math classes. Each of her classes has 12 students. When she found the mean for each class, she was surprised to see that the mean for each class was 80%. However, looking at individual scores, she found that the distribution of scores in each class looked very different.
>
> Write three sample sets of data that have different distributions of quiz scores but have class means of 80%.

## About the Task

When working with quantitative data, students must be able to analyze shape, center, and spread. This task illuminates the idea that different data sets may have the same measure of center even though the data points may look significantly different when analyzing shape and spread. Typically, students are asked to find the mean for a data set. But in this task, the mean is known. Instead, they have to create the different data sets to match the known mean. They must consider different distributions and verify that the mean for each is 80%.

## Anticipating Student Responses

Some students may select a data set with many or all scores of 80%. Other students may then use trial and error to find other data sets that have a mean

PAUSE AND REFLECT

• How does this task compare to tasks I've used?

• What might my students do in this task?

Visit this book's companion website at
**resources.corwin.com/minethegap/6-8**
for complete, downloadable versions of all tasks.

of 80%. *These students may notice that if you add and subtract the same difference, you will not change the mean. For example, two scores of 80 will result in the same mean as two scores of 70 and 90.* Some students may duplicate pairs or sets of numbers that have a mean of 80%. Still others may consider that the sum of the twelve scores is 960. These students may then subtract from or add up to 960.

**MINING HAZARD**

If students use this add-and-subtract strategy, each of their data sets will be symmetric about the mean. These students may believe that the data will always be symmetric.

NOTES

### Student 1

Student 1 creates one representation with all students scoring 8 of 10 possible points. She manipulates this pairing for the second class. With it, she alternates student scores of 9 out of 10 and scores of 7 out of 10. For the last class, she finds a combination of three scores with a mean of 8 out of 10. She then creates four groups of these scores. Each of her classes produces a mean of 8 out of 10 (80%), but her distributions between scores are not significantly different.

### Student 2

Student 2 makes use of different patterns for each representation. In her first class, she finds pairs of numbers with a mean of 80%. She then adds or subtracts 1 from each pair to find new pairs. She leverages a similar strategy with the second class, though her range is considerably different. She gets more creative with her third class. Yet, in both class 2 and class 3, we find instances of scores greater than 100%.

*What would we want to ask these students? What might we do next?*

### Student 1

We must acknowledge Student 1's strategy. She likely believes that the distributions will look different if the data have different values. While the data are slightly different, the largest range for any data set is just 4 points. We should discuss alternate ways to apply her strategy to generate data sets with larger distributions. We might consider making the points out of 100 and finding other pairs of numbers that have a midpoint at the mean. Student 1 will benefit from exposure to and discussion about other students' approaches to the problem.

### Student 2

Student 2 also relies on subset pairings of 80% to generate the data for her three classes. Her first class reminds us of the value of recording ideas systematically so that we can see patterns and make use of them. We should acknowledge this as a sound strategy for finding one of the solutions and note how it can be leveraged for the other classes. We should inquire about the scores greater than 100%. It's possible that extra credit was given, but that isn't included in the prompt. It's fine that she considered it. But it may be that she was simply creating number pairs without considering the meaning of percent relative to the problem's context.

**TASK 41A:** Mrs. Gittermann gave a quiz to each of her three math classes. Each of her classes has 12 students. When she found the mean for each class, she was surprised to see that the mean for each class was 80%. However, looking at individual scores, she found that the distribution of scores in each class looked very different.

Write three sample sets of data that have different distributions of quiz scores but have class means of 80%.

## Student Work 1

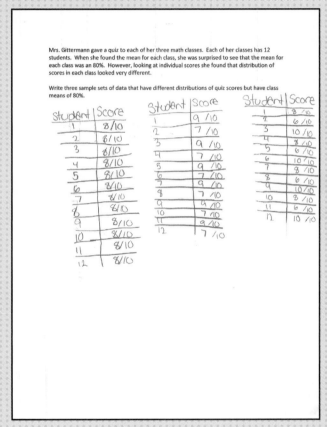

## Student Work 2

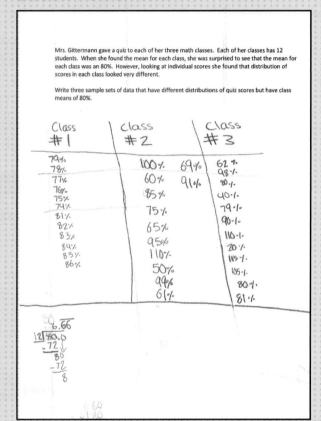

### Student 3

Student 3 reasons about the sum of scores for finding the mean. She notes that the sum of 12 scores of 80 is 960. She then creates different data sets of 12 scores so that each has a sum equal to 960. Student 3 has determined a different strategy for generating a data set for a specific mean. Student 3's strategy may allow for more flexibility when generating individual data values rather than finding individual pairs that have a mean of 80.

### Student 4

Unlike the other students, Student 4 doesn't represent the data set as a list. Instead, she shows a number line and writes the frequency of scores above the score for the first class. For the second class, she states that "6 people got a 75% and 6 people got a 85%"; this is a strategic pairing of data values to average to 80%. And for the third class, she states that "3 people got 50%, 4 people got 100%, 1 person got 90% and 4 people got 80%." Each idea is unique and accurate.

*What would we want to ask these students? What might we do next?*

### Student 3

Student 3's approach is valid and flexible. It's novel. But does it always work? Our next steps with her may be to challenge her to prove that it will and why it will always work. We should also explore the variation of her data sets. Her variation is quite narrow, though her strategy may allow for more flexible findings. We might ask her which data set has the most variation and which has the least variation. We might ask her to represent each with a dot plot or box plot. She, too, offers an approach that will benefit others through discussion and exploration.

### Student 4

Student 4 demonstrates more flexible strategies for generating the data set values. Her sets of numbers are not solely based on pairings. This may signal a more advanced understanding or perception of mean. Student 4 does not explain the strategy she used to generate each of the data sets. Though she doesn't necessarily have to write an explanation, she should justify that the values selected will, in fact, satisfy the prompt. We should also discuss how other representations, including dot plots or frequency tables, could be used to represent the data. We may even consider which may be more appropriate.

**TASK 41A:** Mrs. Gittermann gave a quiz to each of her three math classes. Each of her classes has 12 students. When she found the mean for each class, she was surprised to see that the mean for each class was 80%. However, looking at individual scores, she found that the distribution of scores in each class looked very different.

Write three sample sets of data that have different distributions of quiz scores but have class means of 80%.

## Student Work 3

Mrs. Gittermann gave a quiz to each of her three math classes. Each of her classes has 12 students. When she found the mean for each class, she was surprised to see that the mean for each class was an 80%. However, looking at individual scores she found that distribution of scores in each class looked very different.

Write three sample sets of data that have different distributions of quiz scores but have class means of 80%.

```
                    ⎧ 100        ⎧ 50       ⎧ 80
        960         │ 100        │ 50       │ 80
                    │ 100        │ 50       │ 80
    80              │ 100        │ 50       │ 80
 12)960             │ 100        │ 100      │ 80
                    │ 100    12  │ 100  12  │ 80
               12   ⎨ 100  Scores⎨ 100 Scores⎨ 80
             scores │ 100        │ 100      │ 80
                    │ 100        │ 100      │ 80
                    │ 100        │ 100      │ 80
                    │ 70         │ 100      │ 80
                    ⎩+ 70      + ⎩ 60     + ⎩ 60
                      ___          ___        ___
                      960          960        960
```

## Student Work 4

Mrs. Gittermann gave a quiz to each of her three math classes. Each of her classes has 12 students. When she found the mean for each class, she was surprised to see that the mean for each class was an 80%. However, looking at individual scores she found that distribution of scores in each class looked very different.

Write three sample sets of data that have different distributions of quiz scores but have class means of 80%.

```
   4      6        2
 ├──┼──┼──┼──┼──┼──┤
70% 75% 80 85%90%95%100%
```

6 people got a 75%
6 people got a 85%

3 people got 50%
4 people got 100%
1 person got 90%
4 people got 80%

# OTHER TASKS

- What will count as evidence of understanding?

- What misconceptions might you find?

- What will you do or how will you respond?

Visit this book's companion website at **resources.corwin.com/ minethegap/6-8** for complete, downloadable versions of all tasks.

**TASK 41B:** A clothing designer wants to know how long it takes to sew a new design onto a pair of jeans. Twenty-five tailors at the company are timed as they sew the design. The dot plot shows the time (in minutes) that it takes each person to sew the design.

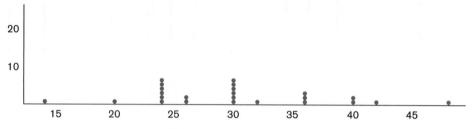

**MINING TIP**

While the mid-range is not always the best measure of center, this value does provide one way to describe "typical" data. Comparing the mid-range to the other measures of center allows students to analyze when some measures of center are better representations of the data.

**The designer wants to know the "typical" time it would take to sew the design. What "typical" time would you suggest? Explain your reasoning.**

Task 41B provides students with an opportunity to explore the concept of center. In it, students must describe a "typical" time it would take to sew the design. Students may identify "typical value" with one of four possible descriptions, each of which could describe a "typical value." Some may select 25 or 30 minutes because this represents the mode for the data. Yet because there are two modes, this is not the best description for a "typical" time. Others may select 30 minutes to represent the mean or the median of the data. *Other students may also select 31 minutes, which represents the mid-range or the center of the interval containing the data.*

**TASK 41C:** A movie studio wanted to see what type of audience was going to see the new romantic comedy in theaters. The box plot shows a sampling of ages of viewers who watched the romantic comedy this week.

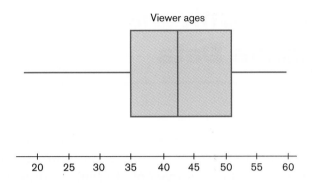

Viewer ages

**Given just this display, what conclusions can you make about the mean and median ages of viewers? Explain your reasoning.**

This task asks students to analyze a box plot and make comparisons about the mean and median. Some students may have difficulty making connections between the shape of the box plot and the measures of center because it only provides summary data for the extremes and quartiles. This task also highlights how skew is represented in a boxplot. Students sometimes think that larger sections represent larger percentages of the data rather than understanding that each quartile represents one-fourth of the data. Some students will have difficulty making a conclusion without having the individual data points as reference. Other students will recognize that the lower bound of this box plot shows a skew to the left indicating that the mean will be affected by this lower extreme. These students will likely and correctly conclude that due to this skew, the mean should be lower than the median.

**TASK 41D:** Create a graphical display in which the mean is greater than the median, but the median is more representative of the data. Explain your reasoning.

Task 41D positions students as creators who select an appropriate display for quantitative data and to create a representation that reflects the shape, center, and spread described. Students must select an appropriate display, which may be a dot plot, histogram, or box plot. They must depict a data set that is skewed to the right, with an outlier as the upper extreme. Some may mistakenly think that having a data set with a large variation guarantees a greater mean than the median. These students are likely not accounting for the median as it needs to be more representative of the data due to an outlier.

# BIG IDEA 42
## Displays of Univariate Quantitative Data

### TASK 42A

Your principal wants to start a "Get Active!" program at your school. He decides to take a survey of students to determine how many minutes, on average, students engage in physical activity each day. The survey results are shown in the histogram.

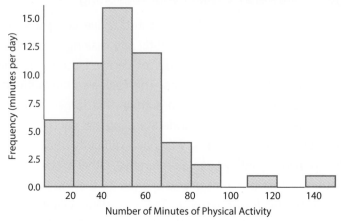

Using what you know about shape, center, and spread, sketch the box plot that would be represented by these data.

### About the Task

Displays of univariate quantitative data provide insights about the shape, center, or spread of the data. Not only do students need to be able to analyze a graphical display, but they also need to be able to translate between different

### PAUSE AND REFLECT

- How does this task compare to tasks I've used?

- What might my students do in this task?

Visit this book's companion website at **resources.corwin.com/minethegap/6-8** for complete, downloadable versions of all tasks.

representations of the same data set. This enables students to understand how the data set helps to define the shape for each display. For this task, students must translate between a histogram and a box plot. Specifically, students need to understand how to translate the clustering and the skew to the right into an accurate box plot representation.

## Anticipating Student Responses

Our students are likely to approach this problem in a few different ways. Some students may try to recreate the data set that is represented by the histogram, then create a box plot based on the data. Some may try to model the shape of the data in a box plot by plotting the extremes and estimating the quartiles. Students may think that the section of the histogram with the greatest frequencies represents the quartiles. Yet as we know, these frequencies do not always translate to the quartile values. Other students may first determine how many data points are in the data set to find where the median and other quartiles are on the frequency table. These students will then use the five-number summary to generate the box plot.

NOTES

## Student 1

Student 1 produces a box plot with minimum 20, lower quartile of 40, median of 60, upper quartile of 80, and maximum 140. His maximum and minimum values are reasonably close to the estimated values. Student 1 shows a box plot that is skewed to the right, matching the right tail of the histogram. However, he doesn't account for the large clustering of data between 30 and 70, which causes the quartile and median to shift to the left.

## Student 2

Student 2 produces a box plot with minimum 0, lower quartile of 20, median of 45, upper quartile of 60 and maximum of 140. He uses the lower bounds of the timebands to estimate the minimum and maximum. He consistently uses this strategy to estimate the quartiles and median. Student 2 also writes the numbers 20, 20, 15, and 80 on the box plot, which appear to represent the distances between numbers in the five-number summary.

## USING EVIDENCE

*What would we want to ask these students? What might we do next?*

## Student 1

We can build on Student 1's understanding. He produces a box plot that clusters to the left and contains most of the data. We should explore other ideas about the data. We might ask him how he could estimate the median on the histogram. This will require examination of the frequencies for each timeband. Of the fifty-three data points, the median happens to fall in the timeband with the largest frequency. This means that the median must be less than 60. Student 1 might then approximate the lower and upper quartiles using the frequencies in the histogram. We can then have him compare his graph to the ideas he shares and discuss how he might adjust his box plot.

## Student 2

Student 2 employs a reasonable strategy to estimate the boxplot. For him, the next step is to label the five-number summary. We can discuss the accuracy of the box plot. We should examine how the provided information in the histogram supports his box plot and what he estimated. The idea to emphasize is that histograms cluster ranges of responses, which may slightly alter the five-number summary. We can have him reinforce his ideas with additional tasks.

**TASK 42A:** Your principal wants to start a "Get Active!" program at your school. He decides to take a survey of students to determine how many minutes, on average, students engage in physical activity each day. The survey results are shown in the histogram.

Using what you know about shape, center, and spread, sketch the box plot that would be represented by these data.

## Student Work 1

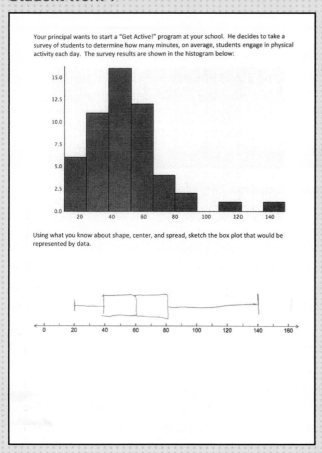

## Student Work 2

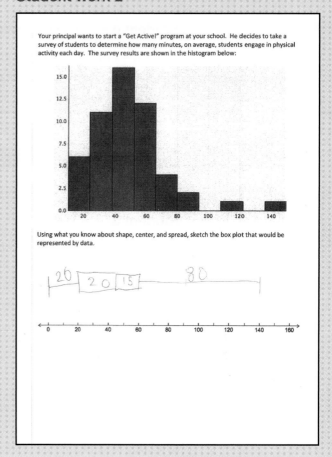

### Student 3

While Student 3 constructs something that resembles a box plot, it is clear that he has significant misconceptions of what a box plot is and how it is constructed. Student 3 plots the minimum at 20, the lower quartile at 80, the upper quartile at 120, and the maximum at 140. There is no median line. Student 3 labels the lower quartile line as "Median" and the upper quartile line as "Mode?" He also finds the difference of 140 and 20 in order to find the mid-range of 80, which he mistakenly labels "median."

### Student 4

Student 4 creates a dot plot rather than a box plot. His example is a fairly accurate representation, but it is unclear if he understands how to interpret information from a histogram to create a boxplot. The scale of his dot plot is not quite accurate.

*What would we want to ask these students? What might we do next?*

### Student 3

For Student 3, we first must correct misunderstandings related to measures of central tendency. We can use a data set that has an outlier. We can have him find the measures of center and compare the values. We should discuss how the outlier affects the mean and median. We then need to revisit what a box plot represents. Is he able to describe it in his own words? We need to determine if he can describe and find the five-number summary. Discuss what each of those critical values represents in the context of the data. He can then work to represent the data as a boxplot. We can observe and discuss the similarities between the data points and the box plot. We may want to work with similar data sets before introducing more varied examples. As we do this, we want him to tell us what different shaped boxplots tell us about the data, particularly about clustering.

### Student 4

We must first ask Student 4 to explain his representation. Is he aware that he hasn't created a box plot? It may be a simple oversight. It may be a bigger challenge. We may have considerable unfinished learning. It appears that he can make sense of a histogram. Does he understand the five-number summary? Is he able to determine these values? Does he understand what a box plot communicates? Does he understand how to create one? Each of these answers has a specific action. As we secure and address answers to these questions, we can revisit this or work with a similar task. His new samples should be evidence of learning.

Your principal wants to start a "Get Active!" program at your school. He decides to take a survey of students to determine how many minutes, on average, students engage in physical activity each day. The survey results are shown in the histogram.

Using what you know about shape, center, and spread, sketch the box plot that would be represented by these data.

## Student Work 3

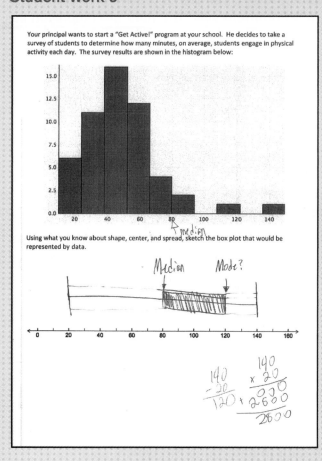

## Student Work 4

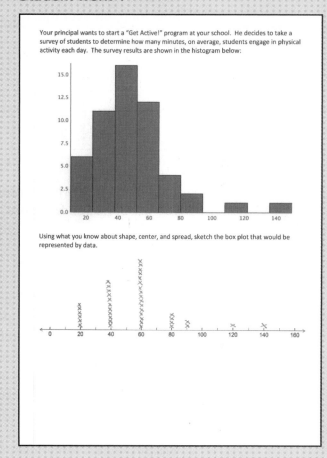

OTHER TASKS

• What will count as evidence of understanding?

• What misconceptions might you find?

• What will you do or how will you respond?

 Visit this book's companion website at **resources.corwin.com/ minethegap/6-8** for complete, downloadable versions of all tasks.

**TASK 42B:** Your school district took a sample of teachers to determine how long, on average, their daily commute was. The data are displayed in the dot plot.

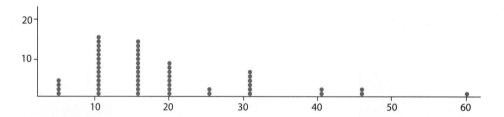

**Based on the shape, center, and spread of the distribution, how would you describe the teachers' daily commute?**

Task 42B requires students to interpret a dot plot to make inferences about the data. Students will need to recognize that the data are clustered between 10 and 20 and that they are skewed to the right due to the outlier at 60. Some may reverse the idea of skew, looking at the data as a whole and seeing a concentration to the left. Students should be able to describe that the daily commute is about 16 minutes. They should note that it can be as small as 4 minutes or as great as 60 minutes. Some students may choose to use the mid-range of about 32 to describe the data. For these students, it will be important to help them understand that the mid-range is not reflective of the majority of the data values.

**TASK 42C:** John collected data to determine the typical salaries of employees at his company. When he analyzed the data, he determined the following:

• **Shape:** The data were skewed to the right.

• **Center:** The center of the data is clustered primarily around $50,000.

• **Spread:** The range of the data is $120,000, with a minimum salary of $30,000. The interquartile range is $25,000.

**Using this analysis,**

• **Sketch a sample dot plot that might represent these data.**

• **Sketch a sample box plot that might represent these data.**

This task asks students to create two different representations for the same data when given information about the shape, center, and spread. Students should find the upper extreme of $150,000. Students should show that this is an outlier, since the data are skewed to the right. Successful students will select lower and upper quartiles that are $25,000 apart, but are much closer to the lower extreme. Students may be able to sketch a general graph but may have difficulty with creating a valid scale that meets the criteria. When supporting these students, we can have them describe what they know from the information provided and how each piece of information translates to a graphical representation.

**TASK 42D:** Antonio was gathering research on the typical heights of middle school students. He collected a sample of 25 boys and 25 girls at his middle school. The box plots are shown. Assuming that this sample is representative of the school population, what conclusions can you make about the heights of girls and boys? Write at least three statements comparing the data.

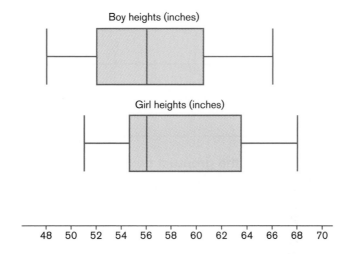

As students build understanding of graphical representations, they learn to compare representations for comparable data sets. In this task, students are asked to write at least three statements comparing the heights of the boys and girls. Some students may confuse the range with a measure of center and thus use this as a justification for why the height of one group is greater than the other. Other students will address the shape, center, and spread of the data. These students will identify that, overall, the girls are taller than the boys since there is a larger minimum, upper quartile, and maximum. They may remark that while the medians for both data sets are the same, the girls' heights are still greater than the boys'. Some students may also conclude that the boys' data are more symmetric, but there is a larger variation in the boys' data than the girls' because the range is larger.

# BIG IDEA 43
## Deviations From the Mean

### TASK 43A

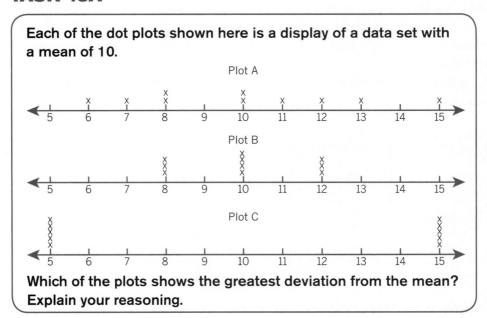

Each of the dot plots shown here is a display of a data set with a mean of 10.

Plot A

Plot B

Plot C

Which of the plots shows the greatest deviation from the mean? Explain your reasoning.

## About the Task

Understanding spread and how variation impacts the mean is central to understanding statistics. As students explore statistical concepts, they learn ways to describe what is "typical" and then analyze other data points to determine whether that data value is significantly atypical by examining deviation from the "typical" value. Tasks for this big idea examine deviations from the mean. Exposure to deviations from the mean is a foundation for later work with statistics. It helps build a basic understanding of standard deviation

## PAUSE AND REFLECT

• How does this task compare to tasks I've used?

• What might my students do in this task?

Visit this book's companion website at
**resources.corwin.com/minethegap/6-8**
for complete, downloadable versions of all tasks.

and how it is used to analyze values within a distribution. In this task, students analyze three dot plots to determine which of the plots shows the greatest deviation from the mean. *The task is unique because students encounter single data sets and find or express the deviation.* In some cases, they look at more than one example.

**MODIFYING THE TASK**

We could modify this task to feature three different representations of data with a mean of 10.

## Anticipating Student Responses

For success, students must consider the mean as a balancing point. Each data value that is to the left of the mean has a weight measured by the distance from the mean in that direction. These weights must be counterbalanced by the data values to the right of the mean. Students may mistakenly think that a dot plot that is symmetric is more balanced than a nonsymmetric distribution. Some may think that plot B is the best choice because it has the most instances of the mean. Other students will analyze the relative position of each of the data values to determine that plot C has the most data values that are farthest from the center.

NOTES

### Student 1

Student 1 incorrectly selects plot A. She describes that this "plot isn't distributed equally to get exactly 10. The others have an even pattern unlike A." Student 1 believes that a distribution must be symmetric for the mean to be in the center. It seems that she does not understand the difference between symmetry and deviation.

### Student 2

Student 2 correctly identifies plot C but is unable to provide a reasonable explanation for why it is correct. Like Student 1, Student 2 is challenged to recognize that 10 is the mean for each of the plots. Student 2 states, "C because the mean will end up to the right." This signals that she thinks that the mean is greater than 10.

*What would we want to ask these students? What might we do next?*

### Student 1

We must first address the misconception that a nonsymmetric distribution will not yield the mean of 10. We might have her calculate the mean with the data points to show that the mean is, in fact, 10. We may also consider examining other nonsymmetrical distributions. We also need to develop understanding of deviation. Physical models such as balance scales may be helpful. Using a scale, we can assign a "weight" to each data value. For plot A, each of the 8s would have a "weight" of 2 on the left side of the balance, whereas the 11 would have a "weight" of 1 to the right of the balance. The total "weights" on each side of the balance will denote the deviations from the mean. Additional work with this tool can help Student 1 determine and compare the deviations of each plot in the task.

### Student 2

Like Student 1, we can have Student 2 calculate the mean for each of the three dot plots to confirm that the mean is 10. We do need to investigate if she understands that there does not need to be a data value at the mean. We might expose her to other data sets to confirm this understanding. We can ask her to describe and compare the ideas of mean and deviation. She too may benefit from physical models of deviation. We should also be sure that additional work to develop her understanding features displays of data in context.

## TASK 43A: Each of the three dot plots is a display of a data set with a mean of 10.

Which of the plots shows the greatest deviation from the mean? Explain your reasoning.

### Student Work 1

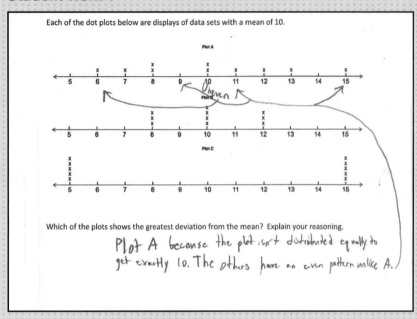

Each of the dot plots below are displays of data sets with a mean of 10.

Which of the plots shows the greatest deviation from the mean? Explain your reasoning.

Plot A because the plot isn't distributed equally to get exactly 10. The others have an even pattern unlike A.

### Student Work 2

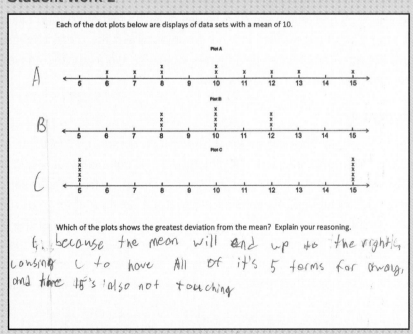

Each of the dot plots below are displays of data sets with a mean of 10.

Which of the plots shows the greatest deviation from the mean? Explain your reasoning.

C, because the mean will end up to the right, causing C to have All of it's 5 terms far away, and there 15's also not touching

### Student 3

Like Student 2, Student 3 also identifies plot C and references that plot C does not have a data value at the mean. She also notes that "Plot A and B has plots on the mean, and plot C has plots on each side." It is not clear whether Student 3 would select plot C if this data set had a value at 10.

### Student 4

Student 4 correctly identifies plot C and is the only student of the four to state that deviation "is the farthest away from the middle." She acknowledges that the mean is 10 and that plot C has the greatest deviation from the mean. From Student 4's explanation, it is not clear whether she has accounted for total deviations from the mean or just how far away the maximum and minimum values are from the mean.

## USING EVIDENCE

*What would we want to ask these students? What might we do next?*

### Student 3

Our first action may be to ask, "If plot C had a value at 10, would this data set still have the greatest deviation from the mean?" Her thoughts may be telling. If she says no, we can address this misconception by showing a distribution with two data values of 9 and two data values of 11 as an example of a data set with a mean of 10. We can use this to explain that while none of the values lie at 10, the mean is 10 and the deviation is relatively small because the data are clustered by the center. We can then separate the values by one increment (two values of 8 and two of 12, then two of 7 and two of 13, etc.). Each time, we can ask her to determine if the mean changes. We can also have her describe how the deviation is changing. We might then have her revisit this task and review her earlier explanation.

### Student 4

We do need to understand what Student 4 means by "farthest away from the middle." We might also ask, "If plot A had one data value at 4 and one data value at 16, would plot A now have the greatest deviation from the mean?" This should shed light on Student 4's consideration of the accumulation of deviations. She too may need other examples to reinforce her conceptualization of the total deviation. Assuming her understanding is sound, we can advance her to consider deviations with distributions represented in different ways. We may also consider having her work with samples that have a greater number of data points as well.

**TASK 43A:** Each of the three dot plots is a display of a data set with a mean of 10.

Which of the plots shows the greatest deviation from the mean? Explain your reasoning.

## Student Work 3

Each of the dot plots below are displays of data sets with a mean of 10.

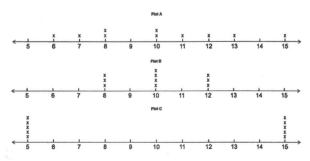

Which of the plots shows the greatest deviation from the mean? Explain your reasoning.

Plot C shows the greatest deviation
from the mean because Plot A and
B has plots on the mean, and
plot C has plots on each side.

## Student Work 4

Each of the dot plots below are displays of data sets with a mean of 10.

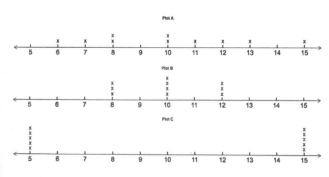

Which of the plots shows the greatest deviation from the mean? Explain your reasoning.

plot C because deviation is the farthest
away from the middle and the mean is 10
so the plot that shows the greatest deviation
from the mean is C.

## OTHER TASKS

- What will count as evidence of understanding?

- What misconceptions might you find?

- What will you do or how will you respond?

Visit this book's companion website at **resources.corwin.com/ minethegap/6-8** for complete, downloadable versions of all tasks.

**TASK 43B:** The data set shown here represents the time (in minutes) it takes a sampling of chefs to prepare a meal. The sample data have a mean of 50. Find the mean absolute deviation. What does this value represent?

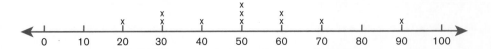

Task 43B is a natural extension of task 43A. Students must apply their understanding of deviations from the mean to find the mean absolute deviation. Students will need to find the total of the individual deviations and then divide by the number of data values (11) to find a mean absolute deviation of about 14.5. This means that the individual meal preparation times vary from the mean by an average of about 14.5 minutes. Often, students are able to calculate the mean absolute deviation but have difficulty describing what this value means. Providing many opportunities with diverse contexts can help develop their descriptions of absolute mean.

**TASK 43C:** Create a dot plot with a mean of 50 and a mean absolute deviation of 20.

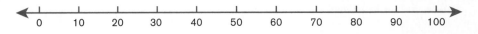

For this task, students apply their understanding of mean absolute deviation to create a dot plot with a mean of 50 and a mean absolute deviation of 20. It may be much more challenging than simply finding an absolute deviation. For success, students will need to understand that, on average, the deviation from the mean is 20. This means that there should be values at least 20 from the mean on each side. Some students may create a dot plot with half the values at 30 and half the values at 70. This is accurate. *It may signal sophisticated, if uninspired, thinking.* Other students may try to disperse the data points. It is essential that they analyze each of the deviations to ensure that the mean absolute deviation is 20.

### ⚠ MINING HAZARD

Our perception of student thinking and student behavior influences our interpretation of their work. Students who simply create data points at 30 and 70 may not necessarily be taking an "easy way out." It may be that their understanding is still developing and this is the *only* way they can complete the task.

**TASK 43D:** Find the mean and the mean absolute deviation for the following data. Create a new dot plot with the same mean and same absolute deviation.

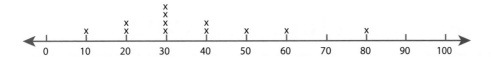

**MODIFYING THE TASK**

*The first prompt is fairly straightforward, designed to measure whether a student is able to find the mean and mean absolute deviation.* Students may look for the "middle number" (median) and deviations from this value rather than calculate the mean. To address this misconception, we might ask if the shape of this distribution indicates that the mean and the median will be the same value. The extension of the task can provide us with insight into how students think about mean and absolute deviation. Some students may change a few data points, whereas others may make significant changes. Still others may create a rather simplistic example that satisfies the prompt. We should also look for students who create a new dot plot with a different number of data points as it may signal even more refined understanding.

To modify this task, have students write a scenario for which this data set is appropriate and then describe what the mean absolute deviation represents in the context of their scenario. This will help to assess whether students understand the meaning of the mean absolute deviation.

NOTES

# BIG IDEA 44
## Bivariate Categorical Data

### TASK 44A

Students from your school were surveyed on whether or not they have a sibling. Use the information below to complete the chart.

- There were 32 people who said they have a sibling.

- There were a total of 28 girls interviewed.

- 11 boys said they have a sibling.

- 50 students were surveyed.

| | Girls | Boys | Total |
|---|---|---|---|
| Has a sibling | | | |
| Does not have a sibling | | | |
| Total | | | |

If this sample is representative of the school, what percent of boys in the school do not have a sibling?

**CONTENT CONNECTION**

Here, students apply their understanding of percents and proportional relationships to solve a problem. This task serves as a nice extension for students as they study ratios and proportional relationships.

### About the Task

Two-way tables provide a means to analyze and compare two categorical variables. *We use two-way tables to estimate likelihoods based on samples and apply proportional reasoning to make predictions about a population.* In this task, students must interpret information to complete the two-way table. They must then analyze the sampling of boys to determine what percent of

### PAUSE AND REFLECT

- How does this task compare to tasks I've used?

- What might my students do in this task?

Visit this book's companion website at
**resources.corwin.com/minethegap/6-8**
for complete, downloadable versions of all tasks.

boys in the school do not have a sibling. Students will need to distinguish between *joint frequencies* (such as the number of boys who do not have a sibling compared to the total students surveyed), *marginal relative frequencies* (such as number of boys compared to the total number of students), and *conditional relative frequencies* (such as boys who do not have a sibling compared to the total number of boys). For this task, students must find a marginal relative frequency.

## Anticipating Student Responses

Students can approach the task in several ways, but each will need to make sense of the relative total for each row or column. Recognizing that 50 individuals have been interviewed is essential for success. Successful students will leverage the combination of these ideas and each new level of information to complete the table. Students may make a range of calculation errors, leading to incorrect values in the two-way table. Some students may incorrectly find a joint frequency by creating the ratio of boys that do not have a sibling to the total number of students surveyed (11 boys of 50 total). The correct finding (11 of 22 boys) is a conditional relative frequency to measure the number of boys who do not have a sibling compared to the total number of boys surveyed.

NOTES

### Student 1

Student 1 uses some of the information to complete the table, but she misses the key information that 50 students were surveyed. She creates other values in the table with a total number of students of 46 rather than 50. This triggers an incorrect answer. Yet her reasoning indicates that she selects the correct table cells to write a fraction comparing boys to boys. Her fraction is not written as a percent.

### Student 2

Student 2 correctly uses the information to complete the two-way table. However, her response indicates that she does not know how to use the information in the table to answer the question. She states, "You can't tell because not all the boys were surveyed." This seems to indicate that she does not understand how to use a sample to make a prediction about the population. It may also indicate that she may think that the same number of girls and boys must be sampled to get an accurate sample.

## USING EVIDENCE

### *What would we want to ask these students? What might we do next?*

### Student 1

We can build on Student 1's ideas. She makes the correct comparison. We may choose to redirect her during her work, or we can choose to let her discover it for herself during whole class debriefing of the task. At that time, we should note errors and oversights that students make. We can provide suggestions for better navigation or, better yet, we can have the class discuss ideas for this. Suggestions might include using a different color pen to denote the information given in the table. Other ideas may suggest rereading the question after a solution is found. Both of these strategies would help Student 1.

### Student 2

We must investigate why Student 2 "can't tell." Her explanation should reveal whether she believes that 28 additional boys (to make 50) need to be surveyed. Or, it may uncover that she does not know how to use a sample to make a prediction based on a population. We can ask what "representative" means as she may not be familiar with the term. We may need to explicitly explain that a representative sample means that it is reasonable to assume that the given information is consistent with the results we would get if all students are represented. With this in place, she can approach the task again. Regardless, we will need to offer additional opportunities to work with sampling.

**TASK 44A:** Students from your school were surveyed on whether or not they have a sibling. Use the information below to complete the chart.

- There were 32 people who said they have a sibling.

- There were a total of 28 girls interviewed.

- 11 boys said they have a sibling.

- 50 students were surveyed.

If this sample is representative of the school, what percent of boys in the school do not have a sibling?

## Student Work 1

Students from your school were surveyed on whether or not they have a sibling. Use the information below to complete the chart.

$$\begin{array}{r} 32 \\ -11 \\ \hline 21 \end{array}$$

- There were 32 people who said they have a sibling.
- There were a total of 28 girls interviewed.
- 11 boys said they have a sibling.
- 50 students were surveyed.

$$\begin{array}{r} 4 \\ \cancel{5}\cancel{0}^{10} \\ -32 \\ \hline 18 \end{array} \qquad \begin{array}{r} 18 \\ -11 \\ \hline 09 \end{array} \qquad \begin{array}{r} 28 \\ -7 \\ \hline 21 \end{array}$$

|  | Girls | Boys | Total |
|---|---|---|---|
| Has a sibling | 21 | 11 | 32 |
| Does not have a sibling | 7 | 7 | 14 |
| Total | 28 | 18 | 46 |

If this sample is representative of the school, what percent of boys in the school do not have a sibling?

$$\frac{7}{18} = 7.1$$

## Student Work 2

Students from your school were surveyed on whether or not they have a sibling. Use the information below to complete the chart.

- There were 32 people who said they have a sibling.
- There were a total of 28 girls interviewed.
- 11 boys said they have a sibling.
- 50 students were surveyed.

|  | Girls | Boys | Total |
|---|---|---|---|
| Has a sibling | 21 | 11 | 32 |
| Does not have a sibling | 7 | 11 | 18 |
| Total | 28 | 22 | 50 |

If this sample is representative of the school, what percent of boys in the school do not have a sibling?

? you can't tell because not all the boys were surveyed.

### Students 3 and 4

Students 3 and 4 make similar errors. Both students use the information provided to complete the table. Student 3 completes the table correctly. Student 4 makes errors with the distribution of girls and the total that have a sibling. Regardless of the mistake, the key table values needed to answer the question are not impacted.

Both students make the mistake of calculating a joint frequency, rather than a conditional relative frequency. Both students are considering the question relative to the total number of students rather than total number of boys. Student 3 also makes the error of having the total in the numerator rather than the denominator. She divides and then adds a percent sign even though the response would be a percent greater than 100.

## USING EVIDENCE

*What would we want to ask these students? What might we do next?*

### Students 3 and 4

Continued work developing the difference between types of frequencies is needed for both Student 3 and Student 4. Before that work begins, we might investigate their perceptions of ratio. Do they recognize that ratio can relate part-to-part or part-to-whole? Their challenges with this task may run deeper than we first consider.

Assuming they do have foundational ratio understanding in place, we can ask about the segment of the population the question specifies. Does it ask about all students or just the boys? We can have them compare their responses to this question with the numerical solution that they provide. Even if they can identify their mistake, we should make use of additional tasks to confirm proficiency with the work.

Student 3's fraction $\left(\frac{50}{11}\right)$ and her result is cause for different concern. It may be indicative of a simple flub. Questions about the work should help us determine the quality of the error. We may have a need for renewed work with percents.

**TASK 44A:** Students from your school were surveyed on whether or not they have a sibling. Use the information below to complete the chart.

- There were 32 people who said they have a sibling.

- There were a total of 28 girls interviewed.

- 11 boys said they have a sibling.

- 50 students were surveyed.

If this sample is representative of the school, what percent of boys in the school do not have a sibling?

## Student Work 3

Students from your school were surveyed on whether or not they have a sibling. Use the information below to complete the chart.

- There were 32 people who said they have a sibling.
- There were a total of 28 girls interviewed.
- 11 boys said they have a sibling.
- 50 students were surveyed.

$$\begin{array}{r} 21 \\ +11 \\ \hline 32 \\ +7 \\ \hline 39 \\ +11 \\ \hline 50 \end{array}$$

|  | Girls | Boys | Total |
|---|---|---|---|
| Has a sibling | 21 | 11 | 32 |
| Does not have a sibling | 7 | 11 | 18 |
| Total | 28 | 22 | 50 |

If this sample is representative of the school, what percent of boys in the school do not have a sibling?

$$\frac{50}{11} = \boxed{4.5\%}$$

## Student Work 4

Students from your school were surveyed on whether or not they have a sibling. Use the information below to complete the chart.

- There were 32 people who said they have a sibling.
- There were a total of 28 girls interviewed.
- 11 boys said they have a sibling.
- 50 students were surveyed.

|  | Girls | Boys | Total |
|---|---|---|---|
| Has a sibling | 13 | 11 | 24 |
| Does not have a sibling | 15 | 11 | 26 |
| Total | 28 | 22 | 50 |

If this sample is representative of the school, what percent of boys in the school do not have a sibling?

22%

$$\frac{11}{50} = \frac{22}{100} \quad (22\% \text{ of boys dont have a sibling})$$

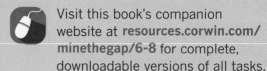

Visit this book's companion website at **resources.corwin.com/minethegap/6-8** for complete, downloadable versions of all tasks.

**TASK 44B:** Riley conducted a survey of her classmates on the following categories. Using the information given, complete the two-way table.

- 40 people said they are right-handed.

- A total of 50 people were surveyed.

|  | Is right-handed | Is left-handed |
|---|---|---|
| Girls | 25 |  |
| Boys |  | 6 |

**Given the sample collected, what percent of girls sampled are left-handed?**

**If there are 250 girls in the school, how many would you expect to be left-handed?**

Similar to task 44A, students must analyze the given information to find the missing quantities in the two-way table. Students must use a conditional relative frequency to determine the solution (40%). Students may find it difficult to determine the percent of girls sampled that are left-handed because the table does not include the marginal frequencies. To assist these students, we might ask them to extend the table to represent the totals for each row and column. This will also help students to check to make sure that the margin total is truly a sum of 50 people surveyed. Students must then use proportional reasoning to estimate the number of girls in the school who are left-handed.

**TASK 44C:** Students in your school were selected at random for a survey to determine if they could ride a bike and ice skate. The results are shown in the table below:

|  | Can ride a bike | Cannot ride a bike | Total |
|---|---|---|---|
| Can ice skate | 25 | 3 | 28 |
| Cannot ice skate | 15 | 7 | 22 |
| Total | 40 | 10 | 50 |

**Based on this sample, what is the probability that a student who rides a bike can also ice skate? If there are 1600 students in your school, how many would you expect to not be able to ride a bike? Using the two-way table, write another question that may be asked, and then find the solution.**

This task requires students to interpret a two-way table for a joint frequency. Students must find that 25 of the 50 students sampled can ride a bike and skate to generate an experimental probability of 50%. Next, students must use proportional reasoning to determine that of the 1600 students in the school, 80% of them (1280 students) can ride a bike. The extension provides the opportunity to create their own problem to solve. Students may create problems requiring use of joint, marginal relative, or conditional relative frequencies. We can use these problems with the class in small or whole group settings. We should look for patterns in the problems they create as it may signal frequencies that we inadvertently feature in our instruction.

**TASK 44D:** Recently, there has been an increase in the number of accidents involving pedestrians wearing earbuds. A research group conducted a random survey of community members to determine how many use earbuds while walking or running. The results are shown in the table below:

|  | Male | Female | Total |
|---|---|---|---|
| Uses earbuds | 37 | 25 | 62 |
| Does not use earbuds | 33 | 25 | 58 |
| Total | 70 | 50 | 120 |

**What are two different conclusions that the researchers could make based on the data? Explain your reasoning.**

Task 44D requires students to analyze a two-way table and draw conclusions. This open-ended task provides an opportunity for students to compare male and female habits with wearing earbuds. *Some students may inaccurately use raw numbers to state that more males use earbuds than females.* Proficient students will recognize that a disproportionate number of males were surveyed, so to compare, they must compare the conditional relative frequencies. These students will find that slightly less than 50% of males use earbuds and exactly 50% of females use earbuds. Students may also choose to compare the total earbud users to nonusers by comparing the marginal relative frequencies.

 **MINING HAZARD**

Students who compare raw totals are likely not conceptualizing relative frequencies. It may be helpful to have students create a graphical representation for the data, such as a stacked histogram or a circle graph, so they can see relative comparisons of earbud use for males and for females.

# BIG IDEA 45
## Bivariate Quantitative Data

### TASK 45A

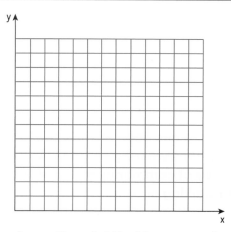

**Construct a sample scatter plot that has a weak negative correlation. Identify a scenario that might be represented with your scatter plot.**

## About the Task

Scatter plots serve as a means to analyze patterns among bivariate quantitative data. In the middle grades, students look for a potential association between two quantitative variables. As students continue to study statistics, they will learn more sophisticated means to analyze the strength of the relationship between two variables. For now, they learn simple techniques for assessing the strength of the correlation. In this task, students must create a scatter plot to model a weak negative correlation between two variables and identify a scenario that might be represented by it. It is a unique task because often our

## PAUSE AND REFLECT

• How does this task compare to tasks I've used?

• What might my students do in this task?

Visit this book's companion website at
**resources.corwin.com/minethegap/6-8**
for complete, downloadable versions of all tasks.

students are presented with scatter plots to interpret or data to plot. In these cases, they may not necessarily make deep meaning of representation or relationships within it.

## Anticipating Student Responses

We might expect students to show four different relationships: weak and strong positive correlations as well as weak and strong negative correlations. Some may create a strong negative correlation, confusing the ideas of strong and weak. Others may confuse these while applying them to a positive correlation. Students may inadvertently create a positive correlation due to the values and context of the situation they create. Successful students will show that a negative correlation exists when one variable decreases as the independent variable increases. These students must demonstrate understanding of the strength of the correlation. *Some students may incorrectly draw a curve to denote a weak correlation.*

**MINING HAZARD**

**Students drawing a curve may select this representation because they have only been exposed to linear models. It is important to show students that data may be modeled by functions other than lines.**

NOTES

### Student 1

Student 1 creates a scatter plot with points that indicate a weak positive correlation rather than a weak negative correlation. He draws a line with a positive slope to represent the expected sales. He assumes that the negative correlation derives from the data points appearing below the expected sales rather than simply underperforming sales.

### Student 2

Student 2 has produced a scatter plot with little to no correlation at all. He provides an amusing scenario for the scatter plot and highlights one data point as an "extra credit person." *The flaw with his scenario in general is that it shows that the longer a person studies, the worse he or she will perform on the quiz.*

**MINING TIP**

We can appreciate our students' humor and clever thinking. However, it is important to reinforce likely outcomes for familiar, real-world situations. Attempts at humor may interfere with our students' understanding or our ability to recognize their understanding.

*What would we want to ask these students? What might we do next?*

### Student 1

We should first talk with Student 1 about his understanding of positive and negative correlation. We can charge him with creating scatter plots to show these correlations. We might first work with context before moving to plots without context. We might have him apply his scenario to one of our decontextualized plots. We need to reinforce with him that the correlation is not relative to expected sales but rather the actual sales over time. He will also benefit from additional opportunities to compare and contrast examples of four types of these correlations.

### Student 2

Student 2 may misunderstand that the "scatter" of the plots indicates both weak and negative correlation. Like Student 1, it is important to first determine Student 2's understanding of positive and negative correlation. We might have him describe what it means for something to have negative correlation and have him show a scatter plot with strong negative correlation. We can have him describe how negative correlation is different. In both cases, we could have him provide an example, or we could provide ready-made examples for him to identify. He could then reconsider his context as well as the correlation that he shows.

## Student Work 1

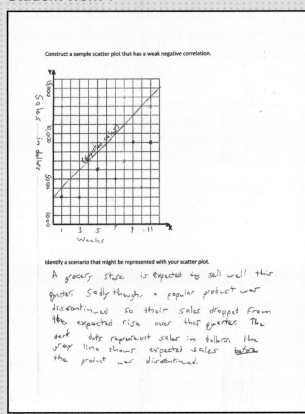

Construct a sample scatter plot that has a weak negative correlation.

Identify a scenario that might be represented with your scatter plot.

A grocery store is expected to sell well this quarter. Sadly though, a popular product was discontinued so their sales dropped from the expected rise over this quarter. The dark dots represent sales in dollars. The gray line shows expected sales before the product was discontinued.

## Student Work 2

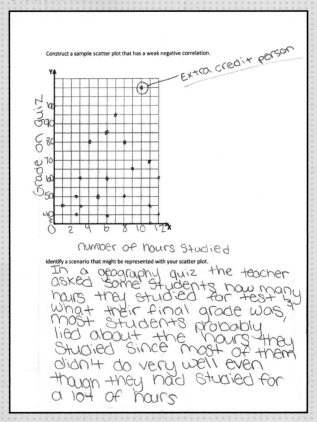

Construct a sample scatter plot that has a weak negative correlation.

Identify a scenario that might be represented with your scatter plot.

In a geography quiz the teacher asked some students how many hours they studied for test & what their final grade was, most students probably lied about the hours they studied since most of them didn't do very well even though they had studied for a lot of hours

### Student 3

Student 3 accurately produces a scatter plot with a weak negative correlation. He is also able to draw a line of best fit that passes through the center of the data. He has, however, created a scenario that does not make sense in the context of the scatter plot drawn. While the scenario selected would show a decrease of water over time, we would not expect data points to be higher over time. Thus, the scenario selected would either be a perfect line or a curve, but the water level should not be rising and falling.

### Student 4

Like Student 3, Student 4 accurately constructs a scatter plot with a weak negative correlation and a line of best fit passing through the center of the data. Student 4 creates a scenario about lemonade, noting that the amount of lemonade sold depends on the weather. While his scenario would show a higher variability in sales, weather conditions are not numerical data to be represented on the *x*-axis. He doesn't provide labels for either of the axes, which likely contributes to his error.

*What would we want to ask these students? What might we do next?*

### Student 3

We should be sure to acknowledge that Student 3's scatter plot points and line of best fit accurately depict a scatter plot with weak negative correlation. To address his scenario issue, we can talk with him about empty water from containers and ask if he has drained a pool specifically. We might have him create a table of values for the scatter plot data, making sure to list the data in ascending order based on the time. Then we can have him examine the *y*-values and describe what the change in these values means in the context of the problem. Once he realizes that some of the water level is rising, have Student 3 adjust his scenario to be a better reflection of the data. Student 3's work is a good example of why student discussion is so valuable. It is likely that the idea of the water level increasing at points would be challenged by his classmates.

**MINING HAZARD**

Understanding concepts and how they relate to real-world situations is essential. Both students show that it is just as important to know mathematics as it is to do mathematics. Tasks that have students doing math may not reveal their understanding of how it applies to their world.

### Student 4

*Like Student 3, Student 4's error focuses on the context of the scenario rather than the mathematics content.* Direct him to create a scale and labels. As he begins to number the *x*-axis, he will likely be challenged to accurately translate this to the weather. At best, he might be able to convert this to number of days selling lemonade or number of hours for selling if the weather is changing throughout a day. The dissonance created should cause him to reconsider his scenario. We can ask about it directly if needed. We can use this error as an opportunity to emphasize the importance of labeling axes and scales, as well as selecting quantitative variables to represent the data. Like Student 3, discussion about his work and thinking can help advance his understanding.

## Student Work 3

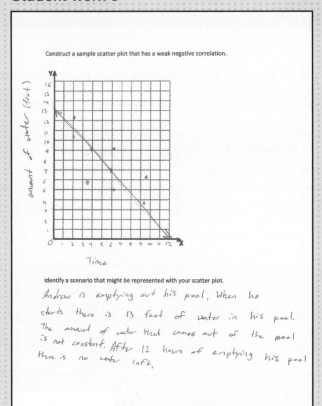

Construct a sample scatter plot that has a weak negative correlation.

Identify a scenario that might be represented with your scatter plot.

Andrew is emptying out his pool. When he starts there is 13 feet of water in his pool. The amount of water that comes out of the pool is not constant. After 12 hours of emptying his pool there is no water left.

## Student Work 4

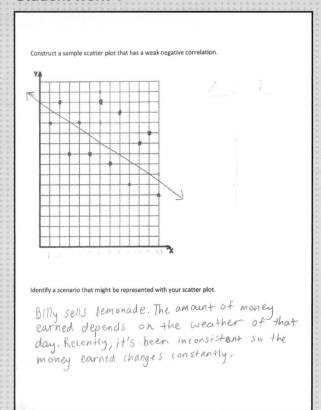

Construct a sample scatter plot that has a weak negative correlation.

Identify a scenario that might be represented with your scatter plot.

Billy sells lemonade. The amount of money earned depends on the weather of that day. Recently, it's been inconsistent so the money earned changes constantly.

## OTHER TASKS

- What will count as evidence of understanding?
- What misconceptions might you find?
- What will you do or how will you respond?

Visit this book's companion website at **resources.corwin.com/ minethegap/6-8** for complete, downloadable versions of all tasks.

**TASK 45B:** Nick wants to enroll his son in a sport. He finds the options shown here in the recreation and parks catalog.

| Sport | Number of Sessions/Lessons | Total Cost (in dollars) |
|---|---|---|
| Soccer | 10 | 80 |
| Football | 8 | 100 |
| Basketball | 12 | 95 |
| Baseball | 9 | 90 |
| Tennis | 15 | 150 |
| Ice Hockey | 14 | 130 |
| Lacrosse | 10 | 90 |

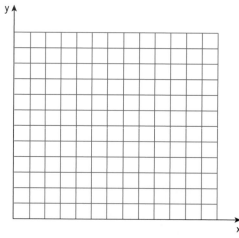

**MODIFYING THE TASK**

We can modify the task to have students estimate a line of best fit or look for correlations in the scatter plot. We might extend the data by changing or adding to the data values.

**Create a scatter plot to compare the cost and number of sessions. Describe the association between the variables.**

This task requires students to accurately draw a scatter plot for the data and describe the correlation between the two variables. *It is an opportunity for students to demonstrate mastery of plotting points in the coordinate plane, selecting and using an appropriate scale, and assessing the relationship between two variables.* Student representations of the data may appear near or far apart because they are selecting the scale. This may be a nice observation and discussion within the class. Some students may plot the total cost on the *x*-axis and plot the number of lessons on the *y*-axis. Successful students will understand that the cost will be based on the number of lessons and that there is a positive correlation between the two variables.

## TASK 45C: Kristen is training for a marathon. She keeps a record of her best times for different runs she has completed in the past few weeks (see downloadable task). Based on her current runs, predict how long it will take her to run 26 miles and how many miles she can run in 40 minutes. Explain how you determined your answers.

Task 45C has students use a given line of best fit to estimate how long it will take a runner to run 26 miles. They have to estimate how many miles Kristen can run in 40 minutes by interpreting or extending the line of best fit. Some students will plot a point on the line of best fit to find the corresponding coordinate to find the value. Some students may create a table drawing conclusions from the values rather than the concept of line of best fit. *Others may choose to estimate their prediction because the data sets do not fit exactly onto the prediction line.* We might extend this task by asking students to describe what the slope of this line represents in the context of this scenario.

**CONTENT CONNECTION**

As students continue to study statistics with bivariate quantitative data, they will need to find the equation to model this prediction line. Students who are fluent with finding an equation of a line should be able to reasonably estimate the equation of this line.

## TASK 45D: Examine the four scatter plots (see downloadable task). Which show a strong correlation between the two variables? Tell how you determined your choice, and write a scenario that would match one of the graphs.

Task 45D asks students to analyze four scatter plots and determine which show a strong correlation between the variables. A misconception may appear as some students will select only one of the graphs. Students may select graph D as they confuse "positive" correlation with "strong" correlation. Some students may choose graph D because they don't believe that a negative correlation can also have a strong correlation. Student scenarios may also provide insight into their interpretation and selection. We should also carefully review each scenario to ensure that the variables selected will have the type of correlation described in the scatter plot.

NOTES

# CHAPTER 8

# WHAT DO WE DO NEXT?

## Take Action . . .

### IMPLEMENT AND COMPARE THE TASKS

There are 180 tasks included in this book. Try them with your students. Use them instructionally. Use them for assessment purposes. Use them to uncover student misconceptions and to determine students who need new opportunities for enrichment and extension. We can compare the sampling of student work in this book with the work of our own students and our colleagues' students. We will likely find many similar results, but we are also likely to find new misconceptions, incomplete understandings, and advanced strategies.

### MODIFY THE TASKS

The tasks presented here are designed for a range of learners and grade bands. Even so, they might not meet the needs of every learner in our class. They might not align to the grade-level content that we teach. Yet, any and all of the tasks presented here can be modified for these differing needs. Ideas for modifying tasks have been offered throughout the book. These offerings are a small subset of the many possible modifications for any of the tasks.

Many of these tasks can be replicated with different numbers or different contexts so that the tasks themselves can become part of the daily routines in our classrooms. Consider this task from Chapter 3.

**Break apart −18 in six different ways using the number bonds below.**

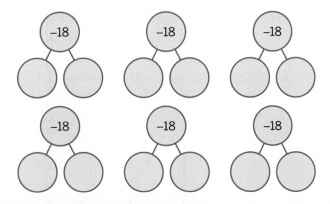

We might change the equations in the prompt to

- Decompose a three-digit integer (−120)

- Decompose an integer into three partials (−18 could be decomposed as −10, −5, and −3)

- Decompose an integer into parts that are both positive and negative (−18 could be decomposed into −20 and 2)

- Compose an expression differently (−18 + −44 could be composed as −10, −40, −8, −2, and −2)

## PLAN COLLABORATIVELY

Working together has much more power than working in isolation. The same is true for planning our mathematics lessons. When we plan together, we learn about others' perspectives, experiences, and mathematical insights. The exchange of ideas helps us reinforce and grow our own ideas. As we plan, it is critically important that we discuss anticipations and misconceptions.

## ANTICIPATE

Anticipation may be one of the most critical, yet overlooked, components of the lesson planning process. In fairness, many of us were not trained to plan lessons with anticipations in mind. Instead, we were trained to think about the materials we would use, the way we would grab our students' attention, and the steps we would take in the lesson. Our planning was teacher centered. Anticipating causes our planning to be more student centered. It helps us plan intentional questions and can help us to make thoughtful and informed "in-the-moment decisions."

## REFLECT ON MISCONCEPTIONS AND INCOMPLETE UNDERSTANDING

Planning collaboratively also sparks conversation about the misconceptions that might reveal themselves during instruction. This happens as teachers share their observations and experiences from previous lessons. It happens as teachers talk about the intent of standards; the progression of concepts; and, most important, their understanding of the mathematics content. As we talk with colleagues about possible misconceptions, we can also begin to strategize about how we can adjust our instruction to address and reteach the misconceptions.

## IMPLEMENT THE TASKS FOR INSTRUCTION

We must make the most of high-quality mathematics tasks. It begins with selection of high-quality mathematics tasks. But that is only part of the equation. It is also critical that we implement them well. Implementing tasks well means that we are facilitators of learning rather than purveyors of knowledge and procedure. It is important that we launch tasks and then get out of the way. This can be tremendously challenging because we can feel that we have less control. An uncomplicated approach to task-based instruction is launching or setting up the task, followed by task engagement and closing with discussion and debriefing.

## LAUNCH, ENGAGE, DEBRIEF

## Launch

We can launch tasks by creating context for the problem. We might read or tell a short story. We might project a captivating image or video that captures the context of the upcoming problem. We might connect the mathematics in the prompt to prior learning. During the launch, we want to provoke our students' excitement for the task and mathematics in general. We also want to confirm that they clearly understand the directions and questions within the task.

## Engage

Our students can engage in these tasks as individuals, as partners, or in small groups. As students work with the mathematics, we must circulate to observe their progress and listen to their discussion. We will have to make decisions about the order in which students or groups will share. We might consider assigning the order of discussion by giving groups numbers on sticky notes as we circulate.

As we circulate, we should consider how student ideas are similar and different. We should consider how representations and strategies are connected. We must also consider how strategies are connected but further refined from group to group.

During this stage, we also may find students struggling with the mathematics concepts or the task in general. It is important that we ask questions to focus their thinking rather than funnel it to our solution pathway. It is also important that we nurture productive struggle while not interfering with their struggle.

## Debrief

Our third stage is to have a class debriefing with the task. At this time, our facilitation is most apparent. We ask students to present their representations, strategies, and arguments. We make use of posters and document cameras to display drawings, diagrams, and calculations. We validate our students' logic. We praise them for their effort as they struggled. We encourage them to look for connections between their ideas and their classmates. We ask them if they agree. We ask questions to focus and move the conversation. Questions we might ask include the following:

- How did you know _____?

- Why did you _____?

- How is your thinking similar to _____?

- How is your strategy different from _____?

Occasionally, our students' strategies and representations may fall short of the messages we wanted to reinforce or the concepts we wanted to establish. In these cases, we can introduce our thoughts and ideas about the prompt. When we do this, it is important that we avoid insinuation of a preferred approach or "correct" strategy. It's also important that we provide follow-up opportunities for students to assimilate the new ideas.

### IMPLEMENT THE TASKS FOR ASSESSMENT

Quality tasks can also be used for assessment purposes. When we do this, we have to be thoughtful about what we expect students to do. We must establish what will be evidence of understanding even if incorrect answers are provided. We also have to consider what will constitute full understanding of the content. With quality tasks, it is important that we value reasoning and logic as much as, if not more than, correct answers. We can make use of general or specific rubrics to support assessment of performance.

We should also be aware of the fine line between instruction and assessment. As we use tasks instructionally, we can also gather insight about our students through observation and discussion. We can take notes about student work or statements. We might use clipboards, recording sheets, or even tablets and data collection software.

## DISCUSS THE RESULTS

Discussing the results with our colleagues is just as important as planning collaboratively. As we discuss the results, we can improve our understanding of what our students' work says about their understanding of the mathematics. Our conversations with colleagues can also help us plan for next steps and, in some cases, reflect on instructional placements. With this in mind, we can begin to see a cycle of planning, implementing, reflecting, and planning again.

## PURSUE THE DEVELOPMENT OF OUR OWN CONTENT UNDERSTANDING

Recognizing and implementing tasks well is grounded in our understanding of the mathematics content. We can do this by reading mathematics education journals or working closely with math specialists or coaches. We can also do this by reviewing reliable online resources and reviewing briefs, positions, and publications from our professional organizations.

An intimate understanding of the mathematics content enables us to better determine the misconceptions that may occur in our classrooms. It helps us recognize when correct answers are not signposts of full understanding. A deeper understanding of the content helps us understand why these responses occur and, most important, what we can do about them. Our understanding also improves our ability to recognize and select quality mathematics tasks.

## PURSUE NEW TASKS WITH OUR REFINED UNDERSTANDING

Now, we have a new understanding of task quality. The 180 tasks in this book align with the big ideas and significant understandings of middle school mathematics. It should also be noted that the tasks offered here are just some of the possibilities for quality mathematics tasks connected to these concepts. However, we are now poised to transfer our understanding to these and other concepts that we teach.

## MINE THE GAP

Mining the gaps in our students' understanding begins with the selection of tasks that we use in our mathematics classrooms. These quality mathematics tasks allow us to access our students' understanding by going beyond recall of fact or procedure. These tasks provoke misconception and incomplete understanding. Uncovering these enables us to take action on them. Without doing so, student misconceptions and flawed understandings linger. Their faulty logic is practiced and reinforced, and the gap widens.

# REFERENCES AND ADDITIONAL RESOURCES

Ashlock, R. B. (2002). *Error patterns in computation* (6th ed.). New York: Macmillan.

Baroody, A. J., & Dowker, A. (Eds.). (2003). *The development of arithmetic concepts and skills.* Mahwah, NJ: Lawrence Erlbaum.

Bresser, R., & Holtzman, C. (1999). *Developing number sense: Grades 3-6.* Sausalito, CA: Math Solutions.

Bresser, R., Melanese, K., & Sphar, C. (2009). *Supporting English language learners in math class: K–2 and 3–5* (2 vols.). Sausalito, CA: Math Solutions.

Burns, M. (2007). *About teaching mathematics: A K–8 resource.* Sausalito, CA: Math Solutions.

Buschman, L. (2007). *Making sense of mathematics: Children sharing and comparing solutions to challenging problems.* Reston, VA: National Council of Teachers of Mathematics.

Chapin, S. H., & Johnson, A. (2000). *Math matters: Grades K–6: Understanding the math you teach.* Sausalito, CA: Math Solutions.

*Common Core State Standards.* (2010). Chief Council of State School Officers.

Dacey, L., & Lynch, J. B. (2007). *Math for all: Differentiating instruction.* Sausalito, CA: Math Solutions.

Fennell, F. (1993). *Teaching for number sense now!* Reston, VA: National Council of Teachers of Mathematics.

Fennell, F. (1998). *Numbers alive!* Parsippany, NJ: Silver Burdett Ginn.

Fostnot, C. T., & Dolk, M. (2001a). *Young mathematicians at work: Constructing multiplication and division.* Portsmouth, NH: Heinemann.

Fostnot, C. T., & Dolk, M. (2001b). *Young mathematicians at work: Constructing number sense, addition, and subtraction.* Portsmouth, NH: Heinemann.

Fostnot, C. T., & Dolk, M. (2002). *Young mathematicians at work: Constructing fractions, decimals, and percents.* Portsmouth, NH: Heinemann.

Krasa, N., & Shunkwiler, S. (2009). *Number sense and number nonsense: Understanding the challenges of learning math.* Baltimore, MD: Brookes.

Litton, N., & Wickett, M. (2009). *This is only a test: Teaching for mathematical understanding in an age of standardized testing.* Sausalito, CA: Math Solutions.

Mokros, J., Russell, S. J., & Economopoulos, K. (1995). *Beyond arithmetic: Changing mathematics in the elementary classroom.* Palo Alto, CA: Dale Seymour.

National Council of Teachers of Mathematics. (1989). *Curriculum and evaluation standards for school mathematics.* Reston, VA: Author.

National Council of Teachers of Mathematics. (2000). *Principles and standards for school mathematics.* Reston, VA: Author.

National Council of Teachers of Mathematics. (2014). *Principles to actions: Ensuring mathematical success for all.* Reston, VA: Author.

National Research Council. (2001). *Adding it up: Helping children learn mathematics* (J. Kilpatrick, J. Swafford, & B. Findell, Eds.). Washington, DC: National Academies Press.

O'Connell, S., & SanGiovanni, J. (2013). *Putting practice into action: Implementing the Common Core Standards for Mathematical Practice.* Portsmouth, NH: Heinemann.

Piaget, J. (1965). *The child's conception of number.* New York: Norton.

Remillard, J. T., Herbel-Eisenmann, B. A., & Lloyd, G. A. (Eds.). (2009). *Mathematics teachers at work: Connecting curriculum materials and classroom instruction.* New York: Routledge.

Sheffield, L. J., & Cruikshank, D. E. (2005). *Teaching and learning elementary and middle school mathematics* (4th ed.). New York: Wiley.

Small, M. (2009). *Good questions: Great ways to differentiate mathematics instruction.* New York: Teachers College Press.

Smith, M. S., & Stein, M. K. (2011). *5 practices for orchestrating productive mathematics discussions.* Reston, VA: National Council of Teachers of Mathematics.

Stein, M. K., & Lane, S. (1996). Instructional tasks and the development of student capacity to think and reason: An analysis of the relationship between teaching and learning in a reform mathematics project. *Educational Research and Evaluation, 2,* 50–80.

Van de Walle, J. A., Karp, K. S., & Bay-Williams, J. M. (2010). *Elementary and middle school mathematics: Teaching developmentally* (7th ed.). New York: Allyn and Bacon.

Williams, J. B., & Karp, K. (Eds.). (2008). *Growing professionally: Readings from NCTM publications for grades K–8.* Reston, VA: National Council of Teachers of Mathematics.

NOTES

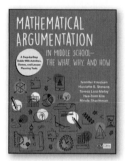

# Supporting Teachers, Empowering Learners

The what, when, and how of teaching practices that evidence shows work best for student learning in mathematics.

John Hattie, Douglas Fisher, Nancy Frey, Linda M. Gojak, Sara Delano Moore, William Mellman

Grades K–12, ISBN: 978-1-5063-6294-6
~~List Price: $36.95~~, Your Price: $29.95

Move the needle on math instruction with these 5 assessment techniques!

Francis (Skip) Fennell, Beth McCord Kobett, Jonathan A. Wray

Grades K–8, ISBN: 978-1-5063-3750-0
~~List Price: $30.95~~, Your Price: $24.95

Students pursue problems they're curious about, not problems they're told to solve!

Gerald Aungst

Grades K–12, ISBN: 978-1-4833-9142-7
~~List Price: $29.95~~, Your Price: $23.95

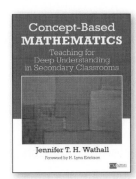

Give students the connections between what they learn and how they do math—and suddenly math makes sense!

Jennifer T. H. Wathall

Grades 6–12, ISBN: 978-1-5063-1494-5
~~List Price: $32.95~~, Your Price: $26.95

Differentiation that shifts your instruction and boosts ALL student learning!

Nanci N. Smith

Grades K–5, ISBN: 978-1-5063-4073-9
Grades 6–12, ISBN: 978-1-5063-4074-6
~~List Price: $36.95~~, Your Price: $29.95 per book

Everything you need to promote mathematical thinking and learning!

Page Keeley, Cheryl Rose Tobey

Grades K–12
Volume 1, ISBN: 978-1-4129-6812-6
Volume 2, ISBN: 978-1-5063-1139-5
~~List Price: $36.95~~, Your Price: $29.95 per book

Corwin educator discount
**20% OFF EVERY DAY!**
★★★
*Quoted price includes discount.

**CM CORWIN MATHEMATICS**

N17828

A SAGE Publishing Company

**CORWIN HAS ONE MISSION:** to enhance education through intentional professional learning.

We build long-term relationships with our authors, educators, clients, and associations who partner with us to develop and continuously improve the best evidence-based practices that establish and support lifelong learning.